Praise for *Kaiāulu*

"Mehana Vaughan's *Kaiāulu* is heartbreaking in its recognition of what has disappeared, and what is slipping away, from her native Hawaiian culture—but at the same time, her book brims with hope, with love, and with reverence for the people who take care of their place down to the last grain of sand. Along with meticulous research, Vaughan offers the reader a portrayal of people she's bound to, through blood and community, and the ways in which tradition and ritual are as essential to human survival as the food we eat."

—Debra Gwartney, author of *Live Through This*

"*Kaiāulu* is an exquisitely written and important book that reflects the beauty of the people whose lives are being so severely damaged by rampant settler colonial capitalism. It is rich with the voices of the Kanaka ʻŌiwi of the Haleleʻa district of Kauaʻi and sweetened with our Native songs and poetry. Its stories make the reader cry out in anger and sorrow at the destruction of gorgeous and fertile land and shores that Kanaka ʻŌiwi have cultivated for a millennium or more. Mehana Blaich Vaughan has issued us a call to action that is impossible to ignore."

—Noenoe K. Silva, author of *The Power of the Steel-tipped Pen: Reconstructing Native Hawaiian Intellectual History*

"*Kaiāulu* brings awareness and hope to all those interested in how we can—together—celebrate, restore, and sustain Earth's rich biocultural diversity. Eloquently and lovingly written, this book provides deep and captivating insights into Hawaiian knowledge. It reflects values and relationships of elders, enriched with Dr. Vaughan's personal experiences and stories—'a lei of stories'—from her family and community. *Kaiāulu* is full of teachings and lessons—some about development, conflict, and loss, and some about responsibility, respect, and renewal. The book is inspiring and a joy to read!"

—Nancy J. Turner, author of *Ancient Pathways, Ancestral Knowledge: Ethnobotany and Ecological Wisdom of Indigenous Peoples of Northwestern North America*

"Mehana Blaich Vaughan has an uncommon eloquence rooted in place. Her native home of Hawai'i speaks to her, holds her close, and instructs her through the voices of her ancestors. She listens to her elders, to fishermen, to the women who live on the coast of Kaua'i, creating the spine of her indigenous community in the midst of tourists, and she weaves these stories into a powerful narrative of sustaining grace on a changing planet. *Kaiāulu* is a book of prayers, an exquisite inquiry into the nature of reciprocity and what it means to be human. Never have we needed the compassionate intelligence of Mehana Blaich Vaughan more. In the tradition of wisdom writers like Robin Wall Kimmerer and the storytelling magic of Louise Erdrich, we see a leader of the next generation on the page and in the world."

—Terry Tempest Williams, Writer-in-Residence at Harvard Divinity School, author of *The Hour of Land*

Kaiāulu GATHERING TIDES

Mehana Blaich Vaughan

Oregon State University Press *Corvallis*

Library of Congress Cataloging-in-Publication Data

Names: Vaughan, Mehana Blaich, author.
Title: Kaiāulu : gathering tides / Mehana Blaich Vaughan.
Description: Corvallis : Oregon State University Press, 2018. | Includes
    bibliographical references and index.
Identifiers: LCCN 2018000980 | ISBN 9780870719226 (original trade paperback : alk. paper)
Subjects: LCSH: Conservation of natural resources—Kauai (Hawaii)
Classification: LCC S932.H3 V38 2018 | DDC 333.7209969/41—dc23
LC record available at https://lccn.loc.gov/2018000980

♾ This paper meets the requirements of ANSI/NISO Z39.48-1992
(Permanence of Paper).

Oregon State University Press
121 The Valley Library
Corvallis OR 97331-4501
541-737-3166 • fax 541-737-3170
www.osupress.oregonstate.edu

All proceeds from the sale of this book will benefit Kīpuka Kuleana, a Hawaiʻi nonprofit dedicated to perpetuating kuleana, ahupuaʻa-based natural resource management and connection to place through protection of cultural landscapes and family lands. Kīpuka Kuleana was founded in 2017 on the island of Kauaʻi. www.kipukakuleana.org

COVER: View of Haleleʻa, Kauaʻi, from Kalihikai as described in opening oli (chant). The mountains of Hanalei at left, Hihimanu, Nāmolokama and Mamalahoa stretch toward Hāʻena in the distance at right, with Mount Makana behind Puʻupoʻā Headland. (Photos by Mike Coots) Cover Art: ʻUpena Kuʻu, Throw Net. (Print by Abigail Romanchak)

For my mother, Beryl Leolani,
for sharing the wonder of books and the joys and *kuleana*
of writing with me and now with my children.

For you—Pikomanawakūpono, Piʻinaʻemalina, and
Anaualeikūpuna, my laughter and my learning,
and for your father, Kilipaki Keliʻiolonoikaʻiminaʻauao,
With all of my love.

In remembrance of my father, Dr. Gary Loomis Blaich,
who rooted us, taught us to love Hawaiʻi and work with joy
to protect places of the heart.

In remembrance of my grandmother,
Tūtū Amelia Ana Kaʻōpua Bailey,
who surrounded us with her lei of warmest love.

And for the many beloved community members of Haleleʻa and
Koʻolau, Kauaʻi, who have helped bring this book to life.

Early morning view of Kīlauea area including Nihokū, Kalihiwai, and the meeting of the *moku* (districts) of Koʻolau at the rocky point in the center of the photo. (Photo by Mike Coots)

Halele'a, Kahale'ala

Halele'a, Fragrant Home

Ku'u haku i ka ua nui o Hanalei
Lei ana o Lū'ia i nā hala o Po'okū
Kū au e hele, ha'alele au 'iā 'oe
'O 'oe kā ka 'uhane i luna o Hīhīmanu
Me he manu lā o Pu'upoā e 'au nei i ke kai
'Akahi au a 'ike i ka mea nui lā he aloha
Aloha 'oe, aloha nō ho'i au ē

My lord in the great rains of Hanalei
Lū'ia is adorned by pandanus trees on the ridge of Po'okū
I stand to go, to leave you
It is you the spirit above Hīhīmanu mountain
Like a bird of Pu'upoā headland swimming in the sea
I have just seen the greatness of aloha
Affection for you, compassion for me

This is a mourning chant offered by Kahale'ala, whose name is also the name of a wind of Kalihiwai, meaning "the fragrant house." Upon returning to Halele'a on Kaua'i after a long journey, Kahale'ala finds his home irreparably changed by the passing of a loved one. Features of the landscape, such as the pouring rains and groves of pandanus (*hala*, to pass), mirror his grief and sense of loss.

(The photo on the front cover of the book shows places described in this *oli* (chant) and the location where Kahale'a may have chanted. This chant is reprinted from the "Mo'olelo [story] of A'ahōaka," published episode by episode in the Hawaiian-language newspaper *Ka Nupepa Kūoko'a*, beginning in 1869, and translated in full by Kamealoha Forrest. This portion of the story appeared in January 1877.)

Contents

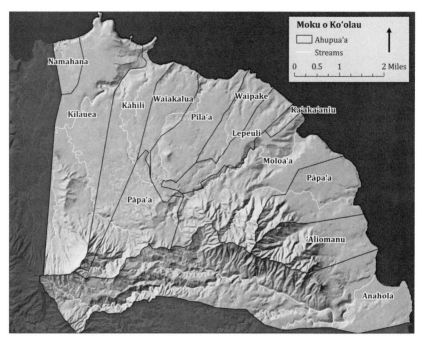

The *moku* (district) of Koʻolau encompasses twelve *ahupuaʻa*, from Anahola in the east to Namahana in the west, including Kīlauea. (Map by Dominique Cordy)

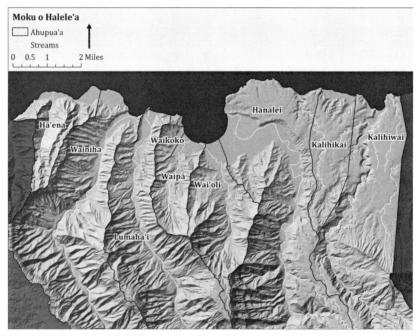

The *moku* (district) of Haleleʻa stretches from Kalihiwai to Hāʻena, encompassing nine *ahupuaʻa*. (Map by Dominique Cordy)

List of Participants

Mahalo to the following people who shared stories for this book over the past twenty years, both through interviews (noted with an *) and through more informal conversations. Those whose names are italicized have passed on during the course of this work. This book is dedicated to their memory.

Akana, Kekoa *
Ako, Valentine *
Alapaʻi, Keliʻi *
Albao, Carol Goo *
Andrade, Carlos
Aniu, Inoa Goo*
Bacon, Makana Aniu
Chandler, Jeff*
Chandler, Kainoa
Chandler, Kaipo *
Chandler, Kapeka *
Chandler, Kapua
Chandler, Kūkaʻilimoku
Chandler, Laʻa
Chandler, Merilee
Chandler, Moku Sr.
Chandler, Pāʻula
Chisholm, Sterling*
Correa, Rachel "Lahela" Chandler
Correa, Lahela Jr.*
Fereira, Mahie
Fereira, Makaleka
Forest, Atta*
Forrest, Devin Kamealoha*

Fu, Ah Meng
Fu, Wendall "Jumbo"
Fu, Kauʻi *
Goo, Betty
Goo, Carol Paik*
Goo, Wendell *
Gowensmith, Debbie*
Harris, Jeremy *
Ham Young, Bobo *
Ham Young, Kalehua *
Hashimoto, Annie*
Hashimoto Goto, Violet *
Haumea, China*
Haumea, Ipo
Haumea, Grandma Lychee
Hermosura, Kaʻimi
Hermosura, Hanalei*
Hoʻomanawanui, Honey-Girl
hoʻomanawanui, kuʻualoha*
Kaʻanana, ʻAnakala Eddie *
Kaʻaumoana, Harry
Kaʻaumoana, Makaʻala *
Kaʻaumoana, Noah *
Kaialoa, Jerry Jr.*

Kahaunaele, Ipo
Kahaunaele, Kainani *
Kahaunaele, Raymond
Kaona, Shorty*
Kinney, Billy *
Kinney, Blue
Kuehu, Nani Paik*
Loo, Audrey*
Lovell, Andrew
Lum, Albert *
Lum, Verdelle*
Mahuiki, Bernard
Mahuiki, Russel*
Mahuiki, Samson*
*Maka, Kapoli**
Maragos, Jim*
Martin, Makana*
Martin, Tamra*
McReynolds, Moana
Mersberg, Kamala*
*Meyer, Sam**
Mission, Bully*
*Miyashiro, Betty**
Molina, Olena *
Nakea, Clifford
Nakea, Robert
Nitta, Keith *
Olanolan, Mike *
Oue, Mama Betty
*Pa, Chauncey**
Pacheco, Gary *
Pacheco, Haunani*
Pa Kam, Annabelle *

Pānui Sadora, Nani
Pānui-Shigeta, Jamie
Paulo Basinga, Maggie*
Paulo, Walter *
Pereira, Charlie*
Pereira, Loke *
Pickett, Tom *
Piena, Moana
Piʻilani, Nancy *
Saffery, Calvin
Saffery, Jenny
*Saffery, Kealoha**
Sloggett, Dick *
Smith, Gary*
Spencer, Jim*
Sproat, David*
Sproat, Linda Akana*
Sproat, Stacy*
Tasaka, Gladys *
Wann, Lei *
Wann, Presley*
Wichman, Chipper*
Wichman, Hauʻoli*
Wilson, Susan *
Winter, Kawika*
Wong, Nathaniel*
Woodward, Blondie*
Woodward, Llwellyn*
Woodward, Nadine *
Woodward, Sam
Yokotake, Naomi*
Youn, Vivian

Prologue

He Lei Aloha (A Lei of Aloha)

Uncle Charlie Pereira of Moloaʻa sewing a throw net. (Photo by Anuhea Taniguchi)

My great-grandfather, Papa James Kaʻōpua, was a throw net fisherman. He and my great-grandmother, Kamila Amina, were both raised in Kohala on Hawaiʻi Island, land of bouldered sea cliffs and relentless Apaʻapaʻa winds. Like many rural Hawaiians at the start of the twentieth century, they moved to the city of Honolulu, on the island of Oʻahu. Together they raised eleven children on Gulick Avenue in Kalihi.[1] Papa fed his family by driving a garbage truck and going *holoholo* (fishing) on the weekends.[2] My Tūtū (grandmother) and her siblings knew that while Papa James was gone fishing with his friend, Uncle Malia, the children were not to argue or talk about what their father was doing, or where. If they did, he would not catch any fish.

One Sunday, Papa was going for *he'e* (octopus) at one of his favorite spots out near Kauhi'īmakaokalani (today known as Crouching Lion), on the Ka'a'awa side of Kahana Bay. The whole family was gathered in Ka'a'awa that day at a friend's home perched on the slope with a nice view of the sea. My Aunty Haunani, oldest of my mom's cousins and first *mo'opuna* (grandchild), was close to Papa. She remembers seeing him out on the reef, pushing the wooden glass-bottom box he used to peer underwater for octopus holes.[3] When she looked back again, the box was floating and Papa was gone.

A week before Papa vanished in 1946, my grandfather, "Grandad" Bob, a naval doctor from West Virginia, had asked Papa for my Tūtū's hand in marriage. My grandparents had met five years earlier, in 1941, at Queen's hospital in Honolulu, where my grandmother was a student nurse and Grandad was a new resident, both of their trainings catapulted into real service when Pearl Harbor was bombed. Papa said yes. My grandparents went on to have five children—four boys and my mother—children who never had the chance to *holoholo* with Papa, their grandfather, or listen to his stories.

Kīlauea sugar lands looking toward the ocean at Kāhili. (Hawai'i State Archives, Photo by 11th Photo Section Air Service U.S.A., 1924)

My mother married my father, who also grew up in Honolulu, when it was far more country, with valleys that still harbored piggeries and farms. They wanted their children to have a similarly rural upbringing, but Honolulu had changed. While visiting friends on the island of Kaua'i in 1977, a few years after Kīlauea's sugar plantation went out of business, my parents found ten acres of former sugar lands set back from the high-end ocean-view-bluff lots. My parents convinced three of my mom's brothers and their families to pool whatever money they could to make the down payment. They all bought the land together, to raise their children and farm. Grandad Bob, who received a single orange in his Christmas stocking each year as a child in West Virginia, gave his children citrus trees to plant. Today, the land that was shoulder-high buffalo grass when they bought it, is canopied by food trees: banana, avocado, breadfruit, grapefruit, tangelo, tangerine, lemon, and orange. My children are learning to harvest from tree limbs bent to the ground with this season's fruit.

The children I grew up with in Kīlauea had harvesting in their blood, descended from vegetable farmers, hunters, and fisher men and women. Their ancestors immigrated to Kīlauea from Japan, China, Portugal, or the Philippines to work on the sugar plantations. Some of the children of these immigrants married Native Hawaiians from the area. These plantation families have fed their children from the mountains and sea of this place across generations, as my Papa fed his children from new places he had to learn. My classmates ate foods my grandmother grew up on—fried *manini*, *lomi 'ō'io*, crack seed *akule*, *aku* bone, dried *he'e*—fish I rarely tasted as a child. I was three when we moved from Honolulu to Kaua'i. Families like mine, newcomers to Kīlauea, were on the cusp of a wave of change, converting former sugar lands, once accessed by the entire plantation community, into private farms and, later, luxury estates. Many of the places I explored growing up, where we rode bikes, built forts, and picked guava, are now gated and off-limits. Most of the kids I went to the local public elementary school with have moved away, priced out by second-home construction and in-migration from the continental United States engulfing sleepy Kīlauea town. Every time I have left, for high school, college, graduate education, work, I have returned to a home more altered. And so I write.

When her children were grown up, my Tūtū, Amelia Ana Ka'ōpua Bailey, Papa's eighth child, took up lei making. Tūtū learned to make flower lei in the *wili* (wrapped) style and showered us with head lei and fragrant strands of *puakenikeni*[4] flowers for every occasion. My grandmother taught us that each lei is distinct to the place where it is made. Through gathering a

My grandmother Amelia Ana Kaʻōpua Bailey making lei *wili*.
(Photo courtesy of Punahou School)

Lei *wili* made for a family *lūʻau*. (Photo by Mike Coots)

variety of fern and flowers from a landscape and weaving them together, lei offer a way to see and know *ʻāina* (land) anew. The northeast coast of Kauaʻi encompasses two rural *moku* (districts): Koʻolau (the trade winds) to the east and Haleleʻa (joyful house) to the west, which meet at Namahana and Kalihiwai, where I grew up. This coast stretches from the Hawaiian Home Lands settlement of Anahola to the former sugar town of Kīlauea, to Princeville resort, to Hanalei Bay, to the end of the highway at Hāʻena. This book is a lei of stories from my home.

My grandmother's four brothers dove the windswept coast of Kaʻaʻawa for a week, looking for their father. They never found Papa's body. One brother later told the family that while swimming along the edge of the reef, searching, he saw *honu* (sea turtles) hovering above the sea floor—over forty, lined up in rows facing the same direction. The family knew they were ancestors who had come to take Papa home.

This story of Papa James's death is what I know about my great-grandfather, along with the *lūʻau* (parties or feasts) he once threw in Kalihi. He invited the neighborhood, and made an *imu* (an underground oven) in the backyard. I also know that after he helped to deliver each of his eleven children at home, he buried their *ʻiewe* (placentas) in that yard and planted fruit trees—guava, persimmon, mango. I've heard that he loved music, liked my grandmother to sing him to sleep. He once bought a player piano for my great-grandmother, who was so angry at his extravagance she ordered the delivery man to take it away. I know Papa was dapper in a *lauhala* (pandanus leaf) hat adorned with fresh lei from the Chinatown lei sellers with whom he shared fish. And perhaps he knows something of me, and maybe that is the reason I have ended up studying the stories of fishing families along the coast, where I grew up.

I have been taught that storytelling, listening to people's experiences then sharing them distilled to their essence, like gathering flowers then making lei, can help people see places and one another in new ways. Shared perspective builds common ground that can change decisions made about the places we love. This is a book about fishing and the responsibilities of fishing families to care for resources on this beautiful coast, now a playground for visitors from all over the world and backdrop for many Hollywood movies.[5] I write to bear witness to the changes I have seen and to honor the families who endure. I share stories shared with me so that the values and relationships that make this community unique can be seen below the surface and carry on. I believe area families' *moʻolelo* (stories) carry lessons

View of Ko'olau, Kaua'i, from Nihokū (Crater Hill) (in the foreground) to Pila'a in the distance. (Photo by author)

View from Kalihikai, looking toward Nīhoku Halele'a continues from here through Hanalei and on to Hā'ena. (Photo by Emily Cadiz)

for how to live in communities shaped by place and how to take better care of the places we love. May you enjoy this lei. May it help you see Koʻolau and Haleleʻa, Kauaʻi, beloved lands to many across the world, in new layered light. May you hear the voices of its loving *kupa ʻāina* (people descended from and enduring in this place).

Pīpī holo kaʻao. May the stories, as always, continue.

A Note About Language

This book draws upon many *ʻōlelo Hawaiʻi* (Hawaiian language) words and concepts fundamental to the stories and ideas shared. Many have no English translation. Hawaiian words are defined in the text the first time they are used and sometimes a second time, when an important word has not been used for a while. Later definitions may differ slightly to include more of the many layers of meaning or *kaona* within Hawaiian terms. To help the reader, there is a full glossary at the end of the book for reference.

Hāʻena reef. (Photo by Mike Coots)

CHAPTER 1

ʻĀina

That Which Feeds

This place will feed you, if you know how to take care of it.
—Makana Martin, Wainiha, 2009

Communities originate in the interdependence of a landscape and its people, a mutuality in which the use of land and waters and the economic, social, and aesthetic choices of residents flourish in balance.
—Beryl Blaich, Kalihiwai, 1986[1]

My alarm goes off at 3:00 a.m. I roll over in the dark, wishing tides had the decency to rise in daylight, so I could too. The day before, when Jeff Chandler told me what time to meet him this morning, he grinned at my expression and said, "The fish don't wait." Knowing he won't either, I place my feet on the woven *lauhala* mat. My feet meet grits of sand.

Where I live on the rural north shore of the island of Kauaʻi, many families continue to catch daily meals from the sea. Fishermen, or *lawaiʻa*, do not invite nonfamily members to go fishing with them and, as a rule, do not share their spots. People do not even say out

Hāʻena *lawaiʻa* (fisherman) Jeff Chandler. (Photo courtesy of Linda Chandler)

loud that they or a loved one are going fishing, lest the fish hear and ruin their luck. Instead, they use the vague term, *holoholo*, cruising around. I will not squander my chance to *holoholo* with a respected fisherman of our coast.

I find the net Uncle Jeff has lent me, one he made years ago for teaching children.[2] It is on the dining table, where I left it after practicing in the dark backyard, hours past when I should have gone to bed. By the last throw, I could unfurl the net, from draped over my arms and shoulders to airborne then open on the grass. It was not a perfect circle, but fried-egg shaped,

1

open enough to fall on at least a few fish. Uncle Jeff calls this motion "a dance." He's asked me to choreograph a throw net hula. I think of the words of my hula teacher's teacher's teacher: "Hula is life." First I have to learn the motions.

I drive the highway to Hāʻena with my brights on, dimming them only once for an oncoming car. I slow to a crawl lest the other driver cross the center line and send me off the cliff to the sea-whipped boulders below. This is the hour of meth addicts on the prowl.

Uncle Jeff's truck is poised to pull out of his driveway, engine running, blinker on, when I pull in. Five boys piled in the camper back are in various stages of half awake. They are his nephews and foster children, my students at our community summer program. Kaʻili, age sixteen, slides from the passenger seat so I can get in and clumps over the tailgate into the bed.

Uncle Jeff turns off the highway onto a dirt road I do not recognize. We bump past a long fence and empty lot that used to belong to his family. We stop facing a towering mahogany gate. Uncle Jeff leans out the window to punch in an entry code. The gate swings open to reveal a porte cochere, lava-rock foundation, second-story porch, and huge picture windows gleaming in the headlights. "You know that TV show?" Uncle Jeff names a nineties sitcom and the actor who starred in the title role. "This is his house. He gives me and my brother the code."

We park in the darkness and fall into line on a path I cannot see through a hedge of *naupaka*, its half-flowers bright in the darkness. As I emerge from the bushes onto the beach, Uncle Jeff is already crossing the sand to the water's edge. I struggle to extract my net from the pillowcase, drape it over my shoulders, and then zip on newly bought *tabi* (reef walkers) without filling them with sand or falling over. The boys walk into the lagoon, six moonlit trails across the still surface that meld into one. I follow, sucking in my breath when the cold water reaches my belly button.

I have not counted on the awkward weight of the net on my shoulders, leads clicking in time with my steps or wind rising before the sun. By the time I reach the reef, gusts are whipping salt spray in my eyes and plucking at the net. We crunch across a nearly dry expanse of coral, following Uncle Jeff, then plunge into a waist-deep channel. I have to bend my knees against the swift current to stay on my feet. Now that it's wet, the net is even heavier. I heave my body back onto dry reef, flopping like a seal. Finding my feet to walk again, eager to catch up, I step where the boys have not. I fall into a hole in the coral and scrape my knee. Kaʻili turns back, "Aunty, I can carry your net for you." I shake my head, but accept his hand to get out of the hole, remembering only later to worry about eels.

Rising Tides of Change

A few minutes later, the sun lifts over the ridge of Hanalei, igniting the ocean and bringing texture to the reef. As gold lights the peaks behind the bay, I forget to care if we ever get a fish. Windows on shore catch the light too. Five-bedroom, three-bath houses sit empty except for caretakers in back cottages and a smattering of well-off tourists staying for $1,500 per night. With each new building foundation dug into these dunes, backhoe blades hit bones. Hawaiians here never lived beachfront, saving the sand to bury their ancestors, Uncle Jeff's ancestors. He is a plaintiff in a case to force a landowner to move construction of his vacation home back two hundred feet, instead of unearthing and moving thirty graves.[3] Uncle Jeff has been arrested twice for trespassing, blocking bulldozers when the landowner proceeded with construction before the suit settled. Coastal luxury homes, owned by investors who live far from Hawaiʻi, sell repeatedly for escalating price tags, five to ten million dollars and up, which drives up neighboring property values. Property taxes are tied to value. At a rate of $6.05 tax per $1,000 property value, a home assessed at three million dollars costs $18,150 per year in property tax.[4] Local families that have lived here for generations can no longer afford their homes, and many have been forced to move away. On one stretch of coast where twenty multigenerational families resided during the 1950s, only two remain today.

For people vacationing or moving here, it must be hard to fathom the sadness and anger longtime residents can carry, as we see loss, and what-used-to-be, everywhere we turn. When I turn back from the houses, Uncle Jeff is far ahead. He turns his outrage into movement, clearing trails, rebuilding rock walls, cutting abandoned fishing nets, and mowing acres of yard at Waipā, where our community summer program is located. At meetings he can be wound tight, with scowling eyebrows and a set jaw, sparking into diatribes that cause fellow community members to cower in our seats. Yet his words reverberate with truth. Uncle Jeff can just as easily erupt into laughter and a generous, flashing smile. As for so many community members, every act, and the anger, is rooted in love for home, and all of us must find our own ways to deal with it.

Communities of Place

Home is a tricky concept in the United States today. People have many places—where they were born, grew up, vacation, have family, lived, and live now. This book speaks to people's need for reconnection, with one another and with places within whose bounds we learn to live. For those in this book, home is specific. When asked where they are from, people of this coast name not Hawai'i, not Kaua'i, nor even Halele'a, the district encompassing the north shore of this island. Instead, they name their *ahupua'a*, or even *'ili*, subdivisions within *ahupua'a*, where their families are from. *Ahupua'a*, one level of Hawaiian land division are "ecologically aligned, and place specific units with access to diverse resources."[5] On the island of Kaua'i, most *ahupua'a* stretch from the mountains to the ocean, encompassing both upland and coastal resources within their boundaries. Each *ahupua'a* has its own winds. According to one wind chant, Wainiha, the *ahupua'a* just next to Hā'ena, has forty; Hā'ena, at least six; Kalihiwai, where I grew up, has three.[6] The names of these many winds, and knowledge of the conditions in which each blows, show the specificity of Hawaiian relationships with place.

Place or land, including the sea, is *'āina*, that which feeds. This word speaks to a relationship with *'āina*, the place that feeds your family, not only physically but spiritually, mentally, and emotionally.[7] For generations, the sea of this coast has fed not only area fishing families, but an entire community, through sharing. The word *kaiāulu* means "community" and contains the words *kai ulu*, the sea at full tide. Community is made up of people connected to and by their relationships with a particular place, as well as the place itself. Community encompasses people who live in, maintain family ties to, advocate over long periods of time for, eat from, or regularly use natural resources in a specific place. Increasingly, today's world is made up of virtual and mobile communities. The stories in this book take readers inside communities of *kupa 'āina* families, those who have become *kupa* (familiar) with *'āina* through generations of living in and eating from a particular place.

Lessons from Living in Place

After twenty years of research, I have come to the conclusion that the character of the relationships Hawaiian fishing families have cultivated with this place is vital beyond these shores. Ancestral values underlying fishing and the struggles of communities to maintain these values in contemporary times offer lessons for achieving a key prerequisite for sustainability: connections with nature based on responsibility, reciprocity, and respect. These lessons emerge not from an untouched native population living harmoniously with their environment, but from communities wrestling with upheavals experienced across the globe.

Today many affluent Americans seek reconnection by searching for the next last "untouched" place. Despite perceptions of Hawai'i as an idyllic island paradise, rural communities on Kaua'i, one of the least populated islands, continue to endure wrenching changes. Diseases introduced with first Western contact in 1778 had killed three-fourths of the Hawaiian population by the time American missionaries arrived in 1820.[8] Immigrant workers from China, Japan, the Philippines, Korea, and Portugal moved to Kaua'i as the subsistence economy shifted to wage labor on sugar plantations and cattle ranches. By the time most of the people I interviewed for this book were born, in the fifty years between 1910 and 1960, Hawaiian families including Uncle Jeff's ancestors had moved from remote valleys to towns and coastal areas close to trade and Western health care. Two tidal waves decimated the coast within a decade, in 1946 and 1957. The first hit early in the morning on April Fools' Day with no warning system in place, claiming ten lives in one small community alone. The ability to survive from the land and ocean, and to share, was essential to recovering from these events. Today, families continue to struggle to hold on to their homes and maintain access to areas that have long sustained them, as wealthy individuals from around the world buy and gate the land. Through all these changes, Hawaiian families, many now Christian and most with Asian and Caucasian ancestors, have continued not only to fish from but also to care for specific coastal areas.

Stretches of reef, each with its own name, are ancestral fishing grounds used by certain families who have fished, eaten from, and watched over them for generations in a system academics describe as local-level resource proprietorship, or a commons.[9] Hawaiian customary and kingdom law limited harvest from inshore fisheries to *ahupua'a* residents. On the north shore of Kaua'i, *lawai'a* continued to respect these exclusive fishing rights of *ahupua'a* families well into the 1970s, three-quarters of a century after the Organic Act proclaimed Hawai'i a US territory and opened most fishing grounds to the public.[10] By harvesting only in small areas, sometimes referred to as a family's "icebox," *lawai'a* got to know fishing spots well, recognize changes, and identify times to rest harvesting from these spots to ensure their abundance.

Kuleana

In Hawaiian, one word, *kuleana*, means rights as well as responsibilities. *Kuleana* is expressed through specific actions or practices that build to create broader impacts when practiced as a community. *Kuleana* is a value. It is also a term used for specific lands. Prior to 1850, *kuleana* were "plots of land given, by the governing *ali'i* [ruling chiefs] of an area or the king himself, to an *'ohana* [family] or an individual as their responsibility without right

'Anakala Eddie Ka'anana cleaning the po'owai (source waters) of the lo'i at Ānuneue. (Photo courtesy of Malia Nobrega-Oliveira)

of ownership."[11] In this definition, responsibility does not stem from owner-ship, but from use and caretaking of land.

When asked the meaning of *kuleana*, beloved elder, *lawai'a*, and taro farmer Anakala Eddie Ka'anana, from Miloli'i on the Big Island of Hawai'i, referred to the lands within a family's care. He explained that a family's work and level of respect in the community, and in turn the rights or authority that family was accorded, were judged largely by the condition of their *ku-leana*.[12] Were these lands *maiau* (well kept) and *momona* (productive and feeding people)? Were their plants, soils, and water healthy? Through his description, and the way he tended the *lo'i kalo* (taro patches) he farmed, Anakala Eddie taught that *kuleana*, all rights and responsibilities, are rooted in the *'āina* itself and how well we care for it.[13]

Kuleana then is more than simply binary rights and responsibilities, but extends to relationships with the land and with other people connected to it. Hawaiian scholar Noelani Goodyear-Ka'ōpua writes that *kuleana* is "au-thority and obligation based in interdependence and community."[14] This book aims to further expand understanding of *kuleana* related to land and resources by investigating authority, obligations, and interdependence within the fishing communities of Hale'e'a and Ko'olau. What responsibil-ities come with being of a place? How can community groups organize to perpetuate *kuleana* to the places they love amid significant encroaching forces? How can communities grow within the bounds of places that sus-tain them, in ways that sustain these places in return?

Hana Paʻa (To Catch, Hold Fast)

It has been an hour and Uncle Jeff is still stalking. I am shivering in my cotton T-shirt, wet up to the neck, wishing I'd worn a long-sleeve rash guard like the boys. They suddenly stop, frozen in place, far back from their uncle, who is crouched low. He is dividing his net the way he taught me when getting ready to throw; three sections, one on each shoulder, *alu* (excess) doubled in his palm. I cannot make out much, though the sky is starting to lighten. Then I see them. Twenty fins, like jaunty America's Cup sails, rise together in the incoming tide. A school of *nenue* (chub), in just enough water to carry them onto the reef, grazes the bed of seaweed.

Years after this morning with Uncle Jeff, when I am studying Hāʻena's fishery for my doctoral research, the oldest living Hāʻena fisherman, Uncle Tommy Hashimoto, will tell me, "I do not look for fish; I meet them when they come home for lunch." Uncle Jeff knew this school of *nenue* would be at this spot on just this tide. He knew, from years of scanning the *limu* (seaweed) beds, the signs of grazing. From years watching the tides change, he learned how much water certain fish need to feed, and how high the tide has to rise to provide that level. He knew how much time we would have before this *ʻāpapa* (section of reef) became submerged and that channel we crossed turned impassable. He knew how to fish the highest reef last. Years of his own observations were piled on years of his father's and grandmother's, cemented from a childhood spent watching them fish.

Just as his nephews watch him now, one wearing a mesh backpack to carry the catch. The boys are still and silent, unlike their teasing, shifting, gangly behavior at our summer program. They know one false step, one shadow cast from the now-rising sun, can scatter the school, ruining both their uncle's chance to catch and the school's chance to feed until the next rising tide. This is a dance. Even I can see it. The fish drop toward the edge of the reef with a receding wave and Uncle Jeff takes a few rapid steps forward in the cover of its foam. When the fish rise with the next surge, he is in range.

Conflicting Coastal Use

Back on shore, I notice the first of the day's tourists, the intrepid ones, starting to peek out of the line of pine trees, getting ready to jump in the water to snorkel. Today, beaches are widely used for recreation rather than as a source of food. According to human use counts that I conducted over a year and a half walking this beach with research assistants in sun and rain, papers blowing from our clipboards, there will be three hundred people here within an hour.[15] Kite surfers will launch from Kanahā Channel, where the turtles feed on seaweed drifting in the current, and rip along this reef. Tourists on snorkeling and diving tours will chuff into the water in groups of twelve at a time. Fifty people will stand on the coral over the course of this day, stop to adjust their masks and kick off with their fins. This bay, whose name, Mākua, means "parent," is known to community members as a hatchery. As children, they were taught not to walk along the shore, lest shadows or footsteps scare baby fish from the shallows into deeper water predators could reach. Today Mākua is visited by over 750,000 of Kaua'i's 1.17 million visitors per year.[16] People pressure on the island has steadily increased as the population of the island has grown over 35 percent since 1990, with visitor arrivals increased 45 percent since 1992 and 25 percent between 2005 and 2015.[17] On an average day, almost twenty-five thousand tourists are on island, a number equal to more than a third of the population of nearly seventy-two thousand island residents.[18]

Community members feel overuse is one of the biggest threats to the health of marine populations on this reef. I have sat in multiple meetings where state resource managers raise questions about how nonextractive recreational use can possibly impact fish populations. Again and again I have heard Uncle Jeff and other fishermen explain that fish feed only on certain tides. Passing kite surfers or paddle boarders appear like predators, causing schools to scatter and miss their chance to feed for the day. After repeated disruptions, the fish will not return to "their houses." "It is like cows," one community member explains, "You want your cows fat and happy, not scared. Fish need to feel comfortable in their homes."[19]

Uncle Jeff's net soars. "Do not throw it too high," he told me, "or the fish have time to see it coming." The net lands in a perfectly symmetrical circle, ten feet across. Then it leaps into the air again as the *nenue* begin to thrash. The boys scramble forward as one. Each loosens a section of monofilament from where it's caught on the coral, moving together as Uncle pulls the net from its center, heavy with fish.

Today's catch is less than these boys' grandparents remember, the fish in them smaller. Fishermen tell me the fish are more skittish. Though large schools still linger offshore, they are no longer breaking up to come into the bays. Is the water on the reef too warm? Is the reef too bustling with tourists? Are the cesspools and septic tanks of the many new vacation rental houses leaching into the bay? Are recreational fishermen catching all the large "cowboy" trophy fish such as *ulua* (jack trevally) that used to split the schools and chase some fish to shore? For Uncle Jeff and the other fishermen of Hā'ena, each issue is a problem in this fishery, and they have grown frustrated with government assigning a separate state agency to each. Fishermen from outside Hā'ena use destructive gear, like lay nets, in ways that were never allowed by area families but that are nonetheless legal under state law. Others use illegal gear but encounter no enforcement. Concerned that their grandchildren will not have fish to catch and eat, families of this community have decided to take action.

Maintaining Connection and Reasserting Values

Uncle Jeff is no conservationist. He once told me, "The only way we know the health of a resource is to eat it." But he is a passionate advocate for Hā'ena, a stronghold for many Hawaiian fishing traditions, the only place I can ever imagine him calling home. Uncle Jeff and other Hā'ena community members are on the front lines of a grassroots movement to bring coastal management back to the local level, in line with ancestral practices, while adapting their fishing to a changing community and climate. In 2006, after years of effort, the people of Hā'ena secured legislation declaring their *ahupua'a* the first permanent community fishery in Hawai'i collaboratively managed for subsistence.[20] A common approach across the Pacific, local-level collaborative management allows communities to work with government to regulate coastal use, setting and enforcing rules for harvest based on ancestral knowledge and relationships with fish, along with real-time observations of their changing rhythms.[21] Nearly a decade later, in the summer of 2015, Hawai'i's governor signed Hā'ena's community rules package into law. How did Hā'ena persevere in this effort through two decades of opposition from both commercial fishermen and the state? What do community members' tenacious efforts to care for a place where most of their families can no longer live teach us about taking care of the places we love?

Ka Hana Lei (Lei Making): A Reflection on *Mo'olelo* and Methods

My people, *kanaka 'ōiwi*, the Native people of Hawai'i, are storytellers who have transmitted the values, lessons, and fabric of our culture in words spoken and, post Western contact, prodigiously written. Like other Indigenous scholars, I use stories or narratives to bind this research together and illustrate the beauty, complexities, and resilience of this place.[22] *Mo'olelo* is the Hawaiian word for both story and history.[23] *Mo'olelo* comes from the word *mo'o*, one meaning of which is succession, to follow one after another, as stories are told from parent to child to grandchild.[24] *Mo'olelo* carry lessons across generations. Hawaiians told teaching stories, not because everyone followed them, but because some did not. In this book, I offer stories of elders and their families, not as history, but as lessons. In teaching us to learn from the past, stories are told to shape the future. As Nigerian storyteller Ben Okri writes, "Nations and peoples are largely the stories they feed themselves."[25]

This book is based on stories shared in over eighty formal and informal interviews conducted over twenty years, sixty recorded and transcribed. These interviews focused on Koʻolau and Haleleʻa between 1910 and the present. This research is also built on a decade of observations of community meetings, participation in community events, coastal use counts, and fishing studies with area fisherpeople. With a team of community research assistants, we have analyzed elder interviews, Hawaiian-language newspapers, *moʻolelo*, land records, maps, and twenty community-rules drafts to govern the coastline of one *ahupuaʻa*. Community advisers, local partners, and young people from the area have helped develop questions, interview their own family members, make sense of the stories we gathered, raise more questions, read drafts of this book, and share. Over ten years, we have

Community research sharing gathering, Hale Hālāwai, Hanalei, August 2009. *Back row, left to right:* Unknown, Kauʻi Quinones, Iams Goodwin, Mauliola Cook, Wendall Fu, Joanna Fu, Kauʻi Fu, Makaʻala Kaʻaumoana, Kawika Winter, Annabelle Pa-Kam, Lilinoe Forrest, Ron McDonald, Lahela Correa Jr., Lahela Chandler Correa, Johanna Ventura, then State Representative Derek Kawakami, Kahanu Peneku, *Front row:* Mehana Vaughan, unknown, Kauahoa Hermosura, interns Teva Noy and Lucía Oliva Hennelly, Pohai Correa, and Louise Sausen. (Photo by author)

held five different gatherings to share findings and get feedback on this work, attended by more than three hundred community members.

Hawaiian culture accords a high degree of secrecy to resource-gathering activities, particularly fishing. Knowledge of fishing practices, and particularly fishing areas, is generally shared only locally, and mostly within families. It would be impossible to conduct this research without being from the area, knowing fishing families and working with them to answer questions and record stories they care about. I am just enough of an insider to be known. I grew up here and taught in community educational programs for years, working with the children and grandchildren of most of my interviewees. I am also just enough of an outsider to be interested and require explanation of that which is obvious to peoples' own 'ohana. Often, I have felt uncomfortable to have other families' stories shared with me so intimately, of a practice of which I am not a practitioner, and of a place to which my 'ohana—having moved here just over thirty-five years ago—remain newcomers. Perhaps it is because the storytellers feel how quickly our community is changing and, like their ancestors writing to Hawaiian newspapers during the nineteenth century, want what they know to be saved.[26] I am not always sure of the reasons people share, and have been hesitant to request people's time to talk story, yet stories have continued to find me.

Recently, I have been asked, by families of kūpuna (elders) I have been blessed to spend time with, to help craft and sometimes deliver eulogies as, one at a time, they have left us. As a respected Maori scholar once stated, "When an elder dies, a library burns."[27] This work and this book have been my effort to ensure that something more than ashes is left, not just for their families, but for a community that loved them, for those who would never otherwise know of them, and for the places they knew like childhood friends.

This book is also based on my own life experiences, raised by my parents on this coast, becoming a parent, and raising my own children here. My family mo'olelo are also shared and woven throughout this lei. Each chapter begins and ends with a story of an event in the life of our community. These mo'olelo illustrate themes within the chapters they enfold, and aim to invite the reader to not just think about but experience the communities of Ko'olau and Halele'a.

Each chapter of this book focuses on a different way of enacting kuleana. Chapter 2, "Hō'ihi: Reciprocity and Respect," is about kuleana to maintain harmonious, mutually respectful, and interdependent familial relationships with the natural world. Chapter 3, "Kahu: Care and Cultivation," describes kuleana to harvest from and cultivate growth within specific areas and the many ways families care for their fishing spots. In chapter 4, "*Konohiki*: Inviting Community Ability and Abundance," I focus on *kuleana* as sharing of harvests, talents, and work to sustain an entire community. Living off the land and sharing builds strong relationships, local-level sufficiency, and resilience in the face of natural disasters, including tidal waves and hurricanes. Chapter 5, "*Kīpuka: Kuleana* to Land," considers the ongoing struggles of Haleleʻa families to keep ancestral lands despite ongoing commodification, rising property taxes, and diminishing access.[28] Though many of these families can no longer live in their ancestral home areas, they work to retain connection and perpetuate *kuleana* in ways that emphasize not ownership but caretaking. Chapter 6, "*Kiaʻi*: Carrying *Kuleana* into Governance," focuses on community efforts to protect and make decisions about natural resources despite conflicts with centralized state management. It deals with the Hāʻena community's twenty-year struggle to develop state law based on local values and practices for coastal use, a model for local communities across Hawaiʻi working to reassert local governance. The conclusion, chapter 7, "*Kaiāulu*: Provisioning Community," ends with voices of the young people of Haleleʻa, the grandchildren and children of community members who share their stories in earlier chapters, extending *kuleana* to future generations. It offers provisions for cultivating reciprocal relationships with place, and for perpetuating community caretaking and responsibility within the context of changing times.

The line between past and present is thin. Descendants are reflections of their ancestors and ancestors live on in their descendants,[29] with stories as a means of connection, providing for both continuity and change. The voices in this book are sharing lessons from their elders, whether they themselves are elders in their eighties or youth in their twenties. This is *moʻokuauhau* (genealogy). This book is an effort to help ensure these *moʻolelo* continue, connected to the place from which they have sprung, and that their lessons are applied in the face of today's challenges in Koʻolau, Haleleʻa, and beyond.

OPPOSITE: This map, hand-drawn by Uncle Gary Smith, longtime Kīlauea community historian, depicts an *iwa* (great frigatebird)-eye view of the western part of Koʻolau, much of which is often referred to as Kīlauea. Uncle Gary draws on multiple resources to depict place-names as well as the winds of each *ahupuaʻa* down the coast. Our community continues to evolve these maps, to adjust and add names as we access archival resources and work together to learn.

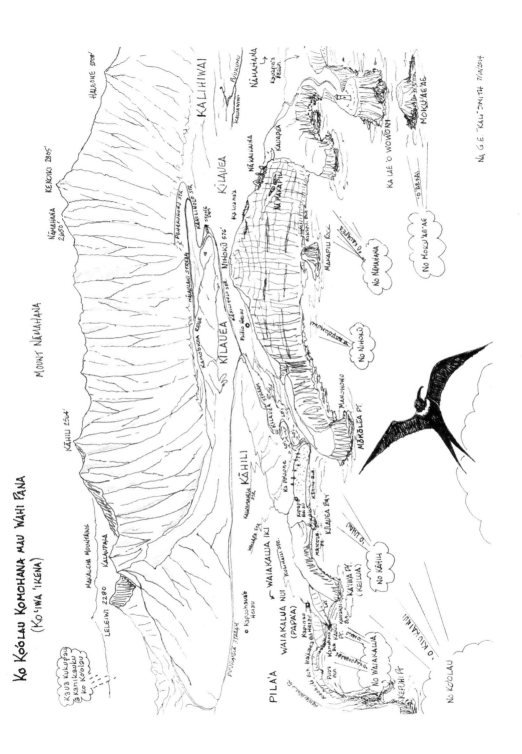

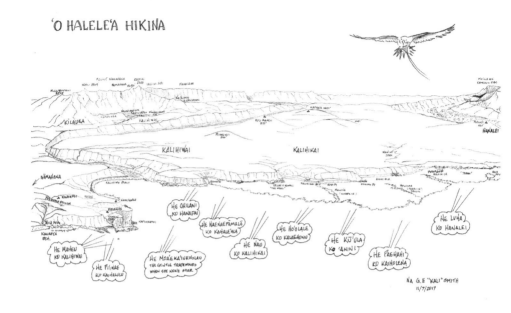

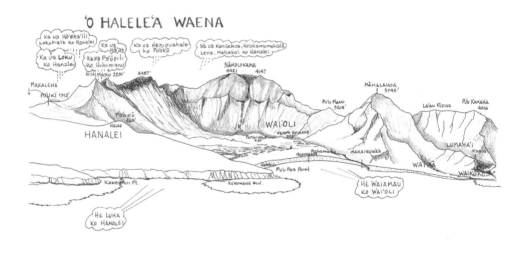

These two maps of Haleleʻa also depict multiple places described in this book, along with place-names and winds along the coast from Kalihiwai to Hanalei Bay, ending at Waikoko. The *ahupuaʻa* of Lumahaʻi, Wainiha and Hāʻena are off to the west (beyond the right edge) of the second map. (Maps by Gary Smith)

Moʻolelo: ʻOhana (Family)

In the summer of 2009, a group of fishermen watched a pile of *akule* (bigeye scad) in Maniniholo Bay for four days. Driving home after finishing fieldwork, I turned the corner at Hale Pohaku and caught the afternoon sunlight glinting off the windshields of their trucks. Ten men stood training their polarized sunglasses on the ocean, watching for the school to come close enough to shore to surround. Each had driven at least half an hour to sit there in silence under the trees each day, setting their lives to the rhythm of the fish.

Akule (Selar crumeno-phthalmus) are a staple species that schools in area bays during the summer months. (Photo by author)

Uncle Bernard Mahuiki joined the group after he closed down the transfer station at 3:30 each afternoon. He was born in Hāʻena with a club foot. As a child, his mother brought him to the beach each morning, burying his foot in the sand to straighten it. His older brother, Uncle Sam, told me, "Bernard was watching fish before he could walk." Hāʻena's oldest fisherman and head *lawaiʻa*, Uncle Tommy Hashimoto, was also there every day among the trucks, offering cold soda from his cooler without taking his eyes from the sea.

After watching for hours one morning, the group saw the color of the pile change. Prized *ʻōʻio* (bonefish) had joined the *akule*. Two men launched the rowboat and slowly unfurled a long expanse of mesh in a wide semicircle encompassing the school. One of the divers swam down to make sure the net wasn't snagged on any rocks or coral before pulling it tighter and moving it toward shore. Before long, he surfaced and jumped into the boat, quaking. He told the head *lawaiʻa*, "There are two huge buggers in the net, swimming around with the fish."

Everyone knew he meant *manō* (sharks). This was odd because, rather than leaping or rushing the sides of the net, the school of fish circled calmly. The head *lawaiʻa* thought about it, and said that since the fish were all right they should proceed with the harvest. He sent divers down to check the net two more times as the boat pulled the catch slowly toward shore. Both times the men came up nodding and shaken. The sharks were still there. When the boat got close to the beach, the head *lawaiʻa* ordered the divers to swim the ends of the net toward shore, then signaled for the assembled crowd to start pulling them in. The first diver started to say something, but the old man shook his head.

Slowly the net began to surface from the sea, fish erupting as they hit shallower water and the net tightened in. Hand over hand, the crowd pulled in the entire net brimming with *ʻōʻio* and *akule*. The *manō* had vanished. The shocked divers protested that they had just seen the sharks in the net; where could they have gone? The old man said nothing. Later, he told his family that the two sharks were his late father and brother. They had come to help their family members to fish, bringing the special blessing of *ʻōʻio* and *akule* to their *ʻohana*. Once they were no longer needed, they disappeared.[1]

NOTE

1.   Story shared by Keliʻi Alapaʻi (2010), in his fifties, the first diver and son-in-law of the head *lawaiʻa*.

CHAPTER 2

Hō'ihi

Reciprocity and Respect

When you talk, the fish can hear and they disappear.
—Charlie Chu, Hā'ena, 2003[1]

Everything is respect.
—Annabelle Pa Kam, Kalihiwai, 2016

The more you share, the more you catch.
—Tommy Hashimoto, Hā'ena, 2010

Hō'ihi, literally "to make sacred," means to treat something with reverence or respect.[2] Fishing stories from the northern coast of Kaua'i describe fisher men's and women's *kuleana* to maintain respectful, reciprocal relationships between people and resources, which together constitute community. Here, people are part of, not separate from, the natural environment.[3] One scholar describes this worldview as a "community of beings . . . in which humans are part of an interacting set of living things."[4] Hawaiian scholar

Uncle Tommy Hashimoto, master Hā'ena *lawai'a*, sharing how to use one of the many nets he has made for *hōlei 'upena* (throw net fishing). (Photo by Ron Vave)

and *kumu* (teacher of) hula Aunty Pua Kanaka'ole Kanahele describes natural resources as "elemental forces, which to us as a people are the deities that sustain our lives."[5] Other descriptions are also based on relationship, such as *nā mea a puni* (everything which surrounds), or *'āina* (that which feeds).[6] Unlike the word "management," which presumes separation and a concept of human power over the environment, these stories illustrate the importance of maintaining mutually respectful and interdependent familial relationships with the natural world.[7]

Mana: Mutually Respectful Relationships

Fishing families in Halele'a describe fishing not simply as harvesting but as building relationships with a particular place, as well as with specific individuals within a given species. When Kaho'ohanohano Pa (1888–1971), a respected head fisherman from the community of Hā'ena, surrounded a school of fish, he always selected one from the net, whispered something to it, and then let it go. According to his daughter, Nancy Pi'ilani, he was naming the fish. "When he went back to catch again, he would call that same fish by name, and it would bring the school back to his nets." Multiple community members recalled Kaho'ohanohano Pa for the close relationship he had with fish in Hā'ena. One elder attributed this *mana* (spiritual power), or ability to attract fish, to Pa's generosity in sharing his catch: "Old man Hanohano, he was so good. He was so generous in sharing the fish. And I've never seen anyone like that; the fish would just come to him. He'd be surrounding a pile at Maniniholo and the schools would just be waiting around the corner at Paweaka, lining up to come to him."[8]

Master *lawai'a* or *kilo* (fish spotters), such as Pa, were keen observers with in-depth knowledge of spawning cycles, weather conditions, moon cycles, and tides.[9] To this day, *kilo* watch schools of fish from the shore or cliffs overlooking bays and direct when and where crews in rowboats drop their nets. In some stories, *kilo* were not just directing the boats based on observed behavior of the fish, but using their *mana* to direct the fish as well. One *kilo* at Kīlauea lighthouse was described standing on Wowoni point high above Kauape'a beach in the 1950s, wearing a red *malo* (loincloth) and chanting schools into the nets below.[10]

As recently as the 1970s, some *lawai'a* kept stones that helped them call or move schools of fish.[11] Describing a well-known fisherman of Halele'a, one of his family members said, "He kept some of the old ways. He had a fishing stone, shrine. They [the family] were Christians yeah. They had a stone. Talk to the fish, because you want to move them to a better place [to catch them]. He asked the fish to move, they move."[12] Schools of fish were said to follow this fisherman as he and his stone traveled around the island.[13] Kahoʻohanohano Pa, described above, kept such a stone. Once, a brother who was jealous of his skill and wanted to catch fish for himself took the stone. "His brother tried to do the same thing but he didn't have that power, that goodness that made the fish come to his net."[14] The power was not in the stones themselves but resided in their keepers' relationship with the natural world: with individual fish within the large schools that roamed the coast, with the schools, as well as with specific *ʻaumākua* or ancestral guardian animals, such as sharks.

Nā ʻAumākua (Ancestors)

For many Hawaiian fishing families, as in the *moʻolelo* preceding this chapter, sharks are *ʻaumākua*, embodiments of ancestors who serve as guardians for their descendants. Multiple species, both marine and terrestrial, were considered *ʻaumākua*, such as *honu* (turtles), manta rays, and *pueo* (owls).[15] *ʻAumākua* were not just any animal of a given species, but particular individuals, many named, who embodied the spirits of recently or long-departed ancestors.[16] The belief that departed relatives become a part of marine ecosystems within their home area encapsulates the idea of families' connection to place. There is no division between humans and nature, only between the living and the departed, with fishing as a way to reconnect.

Sharks are described helping people in reciprocal relationships.[17] In some places, the grandparents of today's fisher men and women fed *ʻaumākua* sharks in rivers or nearshore waters.[18] Fishermen fed sharks a share of their catch, tossing some of the harvest to a particular shark who would in turn help herd fish into nets. One great fisherman of Hāʻena, Paʻitulu, who lived through the turn of the twentieth century, is said to have called a certain shark when he wanted to go fishing along the inaccessible cliffs of the Nā Pali Coast. The shark would carry him on its back down the coast then return him with his catch.[19] Today, while spending time in the ocean, divers and fishermen of families with shark *ʻaumākua* often encounter large sharks, but say they are not afraid, because family members know the sharks are there as guardians.

Families with shark *ʻaumākua* or ancestors have received protection from these animals. One local *ʻohana* gets its surname from an experience, published in a Hawaiian-language newspaper in the mid-nineteenth century, in which family members set off from Kauaʻi in a sailing canoe only to have it capsize at sea. Large sharks came and carried the family members

back to shore at Hanalei Bay. From then on, the family carried the name
Ka'aumoana, meaning to swim the ocean.[20]

Family members of one well-known Kaua'i *lawai'a* (fisherman) recount
him diving to adjust the net while surrounding a pile of fish in a local bay.
He was hit on the head by a lead sinker, knocked out, and sank toward the
bottom. His niece recounts how her father, the fisherman's brother, saw him,
but could not dive deep enough to reach him. Then, the brother saw two
large sharks approach. They swam underneath the sinking, unconscious
man and nudged his body to the surface so that he could breathe.[21] This
family explains the fisherman's survival by saying that sharks are their family *'aumākua*, recounting that the fisherman's grandmother regularly fed a
particular shark in one Kaua'i bay. In these stories, cross-generational relationships bind families of this coast with individual animals of particular
species.

'Ohana (Family)

'Ohana relationships and interactions on land also influence fishing success. Children are cautioned not to argue while a parent is harvesting. As
one fisherman, Bobo Ham Young, explained, "When you go holoholo, everything has to be nice. And the niceness comes from the home."[22] One
kupuna (elder), Uncle Valentine Ako, shared an example from the 1960s, in
which a catch depended on relationships within a fishing family. A school
of *akule* had been in the bay all day, just out of reach of the *konohiki's* (head
fisherman's) net. The *'ohana* was assembled on the beach, waiting to pull
in the school, which drifted with the swells just outside the bay: "The head
fisherman, he had his whole family down there with his net and everything,
but the fish stay outside, and he know somebody stay grumbling."[23]

Sensing something wrong, the *konohiki* gathered his family on the beach
for *ho'oponopono*, a discussion to air and resolve problems, particularly
among family members: "He [got] the whole family together and they had
'ohana [a family meeting] and right there, the whole school comes right up
to the shore. The only thing, you gotta go put the net around them.... I was
told the *akule* has eyes to see and ears to hear and when your *kūpuna* have
problems in the family, the *akule* stays outside, and when you *ho'oponopono*
the *akule* comes in."[24]

Abundant harvests depend on maintaining mutually respectful, harmonious, and interdependent relationships within families, including with resources considered family. These relationships are integral to living in sustainable balance with the natural world.

Mea 'Ola (Sentient Beings)

Fisher relationships are founded on respect for marine life as active participants in fishing. To say the word "fishing" out loud is thought to bring bad luck, because the fish will hear this and avoid being caught. Some elder fishermen on their way to *holoholo* are said to have turned around to head home upon encountering other community members who asked what they were doing, feeling the chances of catching fish were ruined.

Marine species other than fish also choose whether or not to be caught. Fisherwomen of Halele'a and Ko'olau, known for their skills in catching *he'e* (octopus), would walk far out to sea on large barrier reefs, where it is shallow. In this way, some of these women provided most of the protein for their family without ever learning how to swim. One fisherwoman in her sixties from Wanini described going out by herself one day:

> So . . . I see this *he'e* just sprawled out in this tide pool, pretty good sized. . . . So I look at this *he'e* and I say, "You know I really want you but I don't know how I can catch you. Why don't you swim to shore?" And do you know that darn *he'e* swam to shore, and I followed him in and finally when he got to the very end over here, I wooshed him up onto the sand.[25]

Aunty Violet Hashimoto, expert *he'e* fisherwoman and shell-lei maker of Hā'ena with Pi'ina'e Vaughan. (Photo by author)

In another story, marine life helped to select the *konohiki*, who would regulate fishing for an entire *ahupuaʻa*. One woman in her sixties recounted how elders and master fishermen from up and down the coast gathered on a certain moon to select her grandfather as *konohiki*. He was only eight years old at the time.

> Finally one night they decide he's ready. So they make a *pūʻolo* [wrapped bundle] with *ti* leaves. . . . I never did find out what was inside—and they wrapped it all up and his job was to walk on the side of the river where all the rocks are, all the way around to the point. And so he had to *oli* [chant] while he was walking. They gave him this *oli*, so he chanted all the way down. Now this is about midnight, you cannot even see, and they tell him, when you get to this certain rock just stand there and *oli*. And something will come for the *pūʻolo*. So he stands there and chants, scared like anything, his knees were shaking, because it was dark and if he fell he would fall on rocks. And a big shark came up on the flat stone, so he threw his *pūʻolo*. The shark opened his mouth, he threw the *pūʻolo*, the shark closed its mouth, backed up, and left.[26]

The shark appearing and taking the *pūʻolo* was interpreted as a sign that the boy's selection as *konohiki* was acceptable to the ancestors of the place. The young boy became the next area *konohiki* and was later succeeded by his son and grandson, neither of whom had to repeat this approval process.[27]

In these stories, fishing is not just an activity but a relationship of fishermen and women with their catch. It depends not only on skill or knowledge, but also on character and respect. Specific marine species choose whether or not to come to fisher men and women, assist with harvests, and, in one case, confirm selection of local *konohiki* tasked with regulating the fishery. Here there is no separation between people and the natural world. People cannot presume to manage that which sustains them, only to maintain respectful, balanced relationships with all living beings.

Pilina (Relationship) with Place

Relationships with the coast of North Kauaʻi lead to close connections and familial identities shaped not by islands, *moku* (districts), or even entire *ahupuaʻa*, but by specific places and their characteristics, on land as well as in the sea. These connections grow by eating from and sustaining one's family from a particular place over time. One community member from Kalihikai, Blondie Woodward, explained his connection to the area through the taste of its fresh water. "Before, when we were growing up, we had one spring that fed the community—and that was the best tasting water ever."[28] This taste, like the particular taste of the fish and *limu* of that place, would be known, appreciated, and craved only by someone connected to that *ʻāina*.[29] Uncle Blondie further shared, "Before, when I was [living in] Anini, if I go any other place my thirst is not cured until I come home and drink a glass of water. Only then I am satisfied. My thirst is quenched."[30]

Two *kūpuna*, Aunty Loke Pereira of Moloaʻa and Aunty Kalehua Ham Young of Wainiha, talked to me about specific *limu* (seaweed) from their home areas as if describing a box of fine assorted chocolates, each appreciated for its distinct flavor. When I interviewed Aunty Loke, looking out from a high bluff in Koʻolau, she pointed out names of the reef in both directions. She described one species of seaweed that grew on each—that direction, longer, softer, more purple, the other direction, more red and tangy. Adjoining reefs within just one hundred yards of each other not only have different names, but also different delicacies, growing in varied conditions and with distinct tastes.

Aunty Kalehua Ham Young showing Kauʻi Fu and *keiki* of Haleleʻa including Piko Vaughan, Kīwaʻa Hermosura, and Kalālapa Winter how to clean some of the many kinds of *limu* she taught them to gather from a particular area reef. (Photo by author)

Local residents identify these different tastes with home. After naming some of the species her home was known for, one mother in her forties described her associations with the harvest and preparation of these fish: "There was always the smell of the ocean and fish blood. And I like that because it tells me we are gonna have that kind of fish. It helped us to learn how to process fish as well, and how to really enjoy it, and how to eat it, and how to express enjoyment, how to give thanks without saying thank you. How you socialize around food."[31]

Other interviewees describe their sense of home in relation to other means of being fed by a particular place: a certain view, the smell of the air, and the feeling this gives them. As Sam Meyer, in his seventies, said of coming home, "Just the beauty, the freedom. I mean you don't have to love fishing, just the beauty of Anini, the peacefulness of Anini. When I was living in Honolulu and then came home here, I worked in the Honolulu auto industry and [was] just very high strung. I'd hit Kaua'i and just relax. I'd hit Anini beach road, and relax again. Turn that bend to Anini, I'm just totally calm."[32] Llwellyn "Pake" Woodward, in his early forties, describes how he defines home by the air at Kalihikai where he grew up: "When I got married, I had move to Kapa'a side, I had a hard time—just the air on this side alone. It's alive, it's moist. On the other side, once you pass Anahola the air changes. And people don't understand that, they think I'm crazy—If I was to close my eyes and drive over here, I can tell you where I stay."[33] His mother, Nadine Woodward, agrees that when she smells the particular scent of the ocean at Kalihikai, she knows she is home. While today we often think of informational, abstract, and even remotely sensed knowledge of a place, these accounts describe intimate, experiential, and firsthand knowing.

Though Nadine Woodward, her husband, Blondie, and all three of their children grew up in Kalihikai, none of them now has access to their family lands there. However, they return to Kalihikai beach county park for all their family gatherings. The couple speaks in one voice of their love for this area.

> BW: "It was the ambiance of the place."
> NW: "The atmosphere."
> BW: "The people, even though they weren't family—was family."
> NW: "It was like a safe place."
> BW: "Safe haven."

These quotes evidence 'āina, sustaining area 'ohana abundantly, not only physically, through the bounty of its coasts, but emotionally and spiritually as well. Specific places shape people and feed their sense of who they

Nadine Woodward and Blondie Woodward, with their daughter, Kanoe, at a community sharing of research about their home area, Kalihikai (Photo by author, Kalihikai, Kauaʻi, 2015)

are. People consider themselves not only from but *of* a place to which their families have in-depth connections. And the community that makes up any given place consists not just of people but of elements of the natural world, which are also considered family. This concept is described, in the context of other Indigenous communities, as "resource kinship."[34] Successful harvests from the ocean depend not just on knowledge or skill, but also on cultivating reciprocal relationships within families, including with natural resources.

Moʻolelo: *Minamina* (To Grieve for Something Lost, Wasted)

It is July of 2004. I am running Hanalei Bay after a hot day working with the summer program kids at Waipā. Some of those kids are at canoe paddling practice now, warming up. They run by me in the opposite direction, veering from their line of footprints in the sand to give me high fives. I love teaching at Waipā because I see the children everywhere all year long. Where the ocean is still sparkling with gold flecks from the evening sun, quickly sinking over Luluʻupali Ridge, I see a rowboat. Two people inside crouch silhouetted against the dusk. Slowly pulling in their net, they offer a silent reminder that people still feed their families from this bay, named America's top beach in 2009.[1]

Locally, Hanalei is known for *hukilau* (surround net fishing), the *kilo* waiting on the hill at Puʻupoā, watching the black ball of *ʻōpelu* (mackerel scad) or the silver shimmer of *akule* (bigeye scad) moving in. Well into the 1970s, rowboats were loaded with coils of rope hung with *laʻi* (*ti* leaves) dragged over the surface of the water, to herd the fish. Hands *paʻipaʻi* (slapped) the water, rattling the rope and leaves, scaring the fish closer with their shadows. Not until the school was near shore was a smaller net dropped silkily into the sea, to surround the school. Then the *kāhea* (call): "*Huki* [pull] quickly!"

The net was pulled in, hand over hand, its wet weight eased by passing to the next hand. Once the net reached shore, more hands would join to deftly extract the flapping *ʻōpelu* or *akule*, nimbly disentangling their gills. "*ʻEleu ʻeleu* [hurry], no can leave 'em in the net too long." Quickly, surely, never a rush, helpers would toss the bigger fish on the sand for the kids to pick up, and let the small ones go. *Lawa*, enough for everybody.

Forty years ago, surround net *hukilau* happened all summer long, bringing the whole community together to help. However, these days, *hukilau* are less common, carried out a few times a summer in only some of the bays where they used to occur, and hardly ever in busy Hanalei. Participants are fewer, mainly families who regularly fish together, using monofilament nets in place of nylon or *olonā* (a native fiber). Schools are smaller, catches more variable.

As I reach the east end of the bay, I see Uncle Ah Meng Fu's truck on the river side of the pier. He parks there each morning before the sun arrives and each evening as it sets. *Kūpuna* I have interviewed tell me he's the one I should talk to about fishing. He's the one who held the bag and scaled fish for Kahoʻohanohano Pa and the other old-timers when they went *holoholo* down Nā Pali. But times like this, when I do have a chance to talk with him, we mostly sit in silence. Either I'm too shy to ask questions loud enough for him to hear, or he prefers not to answer. Instead of stories, he feeds me cold pickled mango from his cooler. He speaks the recipe, his own, in great detail, the correct amounts to pickle in a mayonnaise jar versus a Foremost ice cream container, steeped to the perfect tangy flavor. I eat "just one more piece" for an hour until the sun begins to set. It is cold. I still have to run back the length of the beach to my car. Uncle says, "Getting dark," he'd better let me go, and gives me a bag of pickled mango to take home.

In the 1960s, visitors riding Kahoʻohanohano Pa's boat to Nā Pali Coast sometimes asked the legendary fisherman if they could record him. Uncle Ah Meng, who drove the boat, told me he cautioned Old Man Kahoʻohanohano to be careful, lest these people take his knowledge for nothing. As I run to my car, I wonder if Uncle Ah Meng ever wishes he had a recording of his teacher. It is too late to ask. I hold Uncle's gift, cool on my palm through the Ziploc.

As the sky darkens, I notice the lights of a fifty-foot sampan that has been floating in the bay day and night for a week, with its fish-finder depth sonar beeping. This big commercial operation, run by Hawaiians from another part of Kauaʻi, has come to Hanalei to surround fish. The sampan's crew never steps foot ashore. A crane angles off the back of the boat to lower the nets, which reach down to the sea floor so that no fish can escape. Whereas local fishing families could *hukilau* three times, feed the whole community, and still leave half a school of *akule* in the bay, these commercial nets surround the entire school.

Community members pole fishing off the pier ask one another, "When they goin' bring 'em in?" The *akule* have been sitting in the net for four days while calls are made up and down the West Coast of the mainland seeking the best price. This evening only half the fish are pulled onto the boat, straight on to ice. The rest of the school is finally released.

That night, *akule* begin to wash in to shore, one at a time. They spin in the shorebreak and are strewn along the beach. First one, then a few, then many. Soon, there are fish at every footprint in the damp cool sand. By dawn, the beach smells like rot between trade winds. Locals stay out of the water,

Hukilau in Hanalei, Kauaʻi, early 1960s. *Back row, left to right:*
Kalehua Ham-Young, Loke Dorian, Howard Kanei, Ah Meng Fu,
Ben Ortiz, Henry Tai-Hook, Lady Haumea, and Unknown. *Front
row:* Rick Ham Young, Kehau Haumea, and Shaine Ham Young.
(Photo courtesy of Billy Kinney)

knowing there will be sharks. The sides of the *akule* belly begin to suck in,
their skin flaking slowly off in the sun. The sampan is long gone.

The next morning's sunrise glints off the windshield of Uncle Ah Meng's
truck. When Uncle Ah Meng, whom no one has seen go farther than the
pier in years, sets watery eyes on the silver *akule* flecking the beach in the
morning sun, he takes a shovel from the back of his truck. He begins to bury
the fish one at a time. Uncle Ah Meng raises his wrinkled face only to sight
the next *akule*, then steps, digs, and stoops to cradle each fish in both hands
before placing it in its grave. He reaches the end of the bay at sunset, feeling
that on this day, he has lived too long. With a sigh of "*Minamina*" (waste), he
turns and trudges, between the burials, the mile and a half back to his truck.

NOTE

1   Associated Press. "Hanalei Bay Named America's Top Beach," US and Canada Travel on
NBC News.com, May 22, 2009 (http://www.nbcnews.com/id/30874693/ns/travel-destination
_travel/t/hanalei-bay-named-americas-top-beach/#.WTuG8IVEyec).

Moʻolelo: *Kai Palaoa* (Whale Sea)

One glance at Hanalei Bay on the morning of July 4, 2004, would tell you it was summer. The water stretched calm, glassy, and sparkling out to sea. Last night's still clouds hovered over the mountains and the trade winds were just beginning to rise from the northeast. Tourists were already plentiful at the beach, with their brightly colored towels, hats, suits, and sun-burnt skin, like wedding flower petals flung the length of the bay, landing in clumps at "Pine Trees" and "The Pavilion."

The sailboats were another sure sign of the season. Twenty or so make their home here each summer, anchored inshore, toward the right side of the bay. I can't say I've ever seen one come or leave. They are just suddenly there. Then one day in the fall—gone. Like the humpback whales that calve in Hawaiʻi each winter and head back to Alaska in spring, these boats seem to be their own migratory species, pulled by Earth's ticking of seasons and tides. We rarely see people on board, and the occasional man rowing a dinghy to shore for groceries seems to be doing his boat's bidding, like a tiny barnacle temporarily dispatched, a mere appendage along for the ride. I wonder what it's like to live on a boat drifting in and out of communities, an island unto yourself?

Toward the left side of the boats, the water appeared strange, shifting, glittering with flashes of black. When my friends and I first took notice we thought it was spinner dolphins, the largest pod we'd ever seen. Upon closer look, we could see that nothing leapt from the sea. The cascading black backs of these animals did not flow in one direction, but drifted listlessly flopping from side to side. These were melon-headed whales, *palaoa*,[1] a deepwater species rarely spotted nearshore in Hawaiʻi except before stranding.[2] That evening we watched the sun set over their backs, both dreading and hoping they'd be gone in the morning.

One scientist quoted in the newspaper said that this species, with their torpedo-shaped bodies and blunt heads, would not be able to find food in the still waters of the bay. He figured the pod of 100–150 animals had not eaten since they'd arrived and would die if they didn't leave soon. Another expert claimed melon-heads came into still bays like Hanalei to calve. Still another related their appearance to recent offshore naval exercises testing use of sonar. No one really seemed to know.

I knew the air was different with their presence. All of Hanalei felt charged with something so out of the ordinary that it must be trying to shake us out of our regular routines, to make us understand . . . something. Why had the *palaoa* chosen to visit? They must be bearing a message, whether blessing or warning, I wasn't sure. So, quietly, almost secretly, I would send prayers to the whales, chant to honor these ancestors. I sent the words across the waters of the bay, waiting each day for their answer.

Setting out on our morning run, my two friends and I noticed that the whales seemed to be getting less active. They drifted near the boats, then down toward the middle of the bay, now back alongside the boats. Onlookers speculated that the whales would soon beach. The area in front of The Pavilion, directly inshore from the whales, buzzed with activity. A trailer backed up on the sand, unloading six plastic scupper kayaks to join a flotilla of pink, orange, and lime-green ones already lined up. Members of the Hanalei Canoe Club, including a large group of kids, were arriving in their bright yellow and green jerseys, carrying paddles. Three canoes were gliding over from the river mouth. More and more people gathered, milling about like the whales.

As we ran, dodging our way through the throng, one friend exclaimed, "It looks like they're going to try to paddle out there!" It certainly did. I shuddered at the thought of all those plastic scuppers careening into the pod, paddles hacking.

"I wish people would just leave the whales alone," said my other friend, as we neared the end of the sand and entered the water to swim back.

Whatever was about to happen, neither of my friends wanted to see it, and they swam quickly ahead of me toward home. I swam slowly, tangled in my own worries. Was that crowd about to descend on the whales? Should we stop them? What if the melon-headed whales did need help? What were they trying to tell us? What were we supposed to do? I couldn't just leave and go home.

Suddenly, I heard a loud metallic sound. A lifeguard on the beach was yelling through a megaphone, "You. Get out of the water. There is no swimming in this area. Yeah, I mean you two. Out." His megaphone was aimed at my friends who'd swum back to the area in front of the crowd. The crowd stopped moving and turned to stare. "We have closed this area," repeated the lifeguard. "Get out of the water. Now." I could see yellow streamers stretched perpendicular to the ocean on either side of the Pavilion, closing the ocean inshore of the pod. I watched from the water as my friends got out of the ocean, the crowd of eyes still on them, one clutching the saggy fabric of her favorite old bikini bottoms. They walked around the yellow tape to where they could reenter the ocean and swim home.

I was going to get yelled at too if I tried to swim any farther along the shore. Ducking my face under the water to clear my head, I heard a high-pitched sound, almost like a whinny, followed by another and some clicking. The whales were talking.

I turned toward the sea and began to swim to the nearest boat. I huddled next to its hull where the lifeguards wouldn't see me, slid slowly around its edge, and then dove underwater. The whale sounds grew louder, as I reached the bottom of the next boat. This boat had people on it, eating breakfast at a table. I waved, then swam to the next vessel, before they could get curious about what I was doing.

I didn't know what I was doing, only that all those people were about to try to save these whales, and I had to understand if the whales thought they needed saving or not. The whales were ancestors. I couldn't know how I felt about anything people might do without tuning into them first. I ducked around the stern of one more boat and looked out to the open bay, all the way across to Waikoko. The water here was deep, deep blue. If I swam down to touch the bottom, I wouldn't have enough air to come back up. About forty whales were straight in front of me, less than twenty-five yards away. I pushed back thoughts that came bubbling up. Did melon-headed whales have teeth like killer whales? Might sharks be gathering near the pod? I clung tight to the boat and stuck my face underwater to listen.

Now, I was in the heart of the sounds, and they reverberated around me. High-pitched vocalizations were interspersed with fast clicking. In such deep ocean, the sounds echoed eerily close. I yanked my head out of the water. From this vantage point, I could see the whales nudging against one another. Each whale faced a different direction, like pick-up sticks bobbing on the current. They rolled through the water, not splashing their flippers. They occasionally opened their blowholes to emit a sound of breath and mist so faint it vanished by the time the holes closed up again. Their fins flopped over in cartoon curlicues, hanging limp.

Meanwhile, the crowd on the beach had stopped milling around and gained focus. Some people spread out along the vegetation line. They were pulling long streams of *pōhuehue* (beach morning glory) vine and weaving them together. Others were pulling the kayaks farther up the sand. Another group was gathered closely together, facing a speaker who was waving his hands authoritatively.

As I watched, the groups of people formed a circle holding hands. They stood still and silent, heads bowed, then lifted their faces to the sea. A sound rose across the water, the first notes of "E Ho Mai," an *oli* asking for wisdom and guidance from above, for understanding. I felt the whales move, and when I turned to look, they were turning too. Each whale swung to face the chanters until all their heads pointed toward shore, tails to the ocean, row after row, dorsal fins straight up in the air, listening.

When the chant ended, the group on the beach broke their circle, and the whales too returned to milling about. But I felt they knew something was about to happen and they were ready. Four canoes began to paddle out toward the whales, a long rope of *pōhuehue* vine stretched between them like the old-timers used in place of rope to herd schools of fish in *hukilau*. Each canoe advanced slowly, still far from the whales. The two canoes on the outside paddled ahead of the others so that the vines stretched out in a crescent. The paddlers were all ages, mostly kids. Their canoe club coach, Hanalei Hermosura, my age and named after this bay, stood in a rowboat. He moved back and forth between the canoes, giving instructions: "Paddle up, slowly, slowly. There, stop. Red canoe, paddle back. No hurry. Judy, steer your canoe a little more toward me. OK. Steady, steady."

No one else spoke, just listened for his instructions. All sat poised, ready to paddle up, freeze, back paddle, hold, and ease the vine rope gently, little by little, around the pod. No boat ever got closer than twenty feet from the whales. No one jumped in to swim with the whales. No paddles reached out to touch them or even splashed. Each boat moved as a unit, slowly, haltingly, parallel parking the vine into place. Soon, the vine encircled the pod, with an opening toward the sea. Gingerly, the canoes began to paddle toward the horizon together, pulling the vine forward—its motion through the water sent ripples to the depths, whispering the way. The pod began to still its restless seething and shift slowly toward sea. The canoes paddled softly, slowly navigating the whales away from the beach. Nudging one another, the whales began to raise their dorsal fins and turn toward the horizon. All facing the same direction once again, the pod started to swim steadily out of the bay.

Still clinging to the molding on the side of the boat, I watched the whales until they were too far away to see from water level, then hoisted myself out of the water for a better view. The canoes followed only a little farther, then stopped, people straining from their seats to see as the pod swam stronger and stronger, gaining rhythm toward the open ocean. Suddenly, one whale jumped, soaring out of the sea, all of us witnessing the torpedo-shaped body for the first time. Two children shouted, their voices joining its landing splash. Then another whale jumped and another, until everyone in the canoes was encouragingly whooping, laughing, crying, waving good-bye.

The next afternoon, I took the ten children at our Waipā community after-school program walking along the bay to tell them about the whales. Two of the ten-year-old girls ran ahead. They started shouting and waving their arms just before they reached the Waikoko rocks, at the spot called Ke Ahu (the altar), which marks the *ahupuaʻa* boundary between Waikoko and Waipā. "Aunty! Aunty come!" one called.

A baby melon-headed whale the size of a fire extinguisher lay on the sand where the children stood. I was struck that something so young could look so old. When they saw it, each child in turn went from chattering away to instantly silent. Three stooped to lift it back into the water, cradling its body, urging it to swim. When they realized it would not, they backed out of the ocean and sat on the sand, holding the whale in their laps. The younger children drew close, knelt to touch its leathery face, traced scars from a net and two cuts on its side. "Aunty, what should we do?"

They wanted to chant. One of the older girls suggested the chant, "*Aloha e nā kūpuna*" (Love for our ancestors). "Because it is our ancestor," she said, "Even though it is only a baby."

After they chanted, I called NOAA's marine mammal response team, who sent a female marine biologist to collect the specimen. The kids wanted to bury the whale, pray, sing, plant its body under a tree where we could come back and visit. The biologist explained that if she took the baby, she could study it, look into its brain to see what made it beach itself and maybe keep this from happening again.[3] Reluctantly, the children handed her the whale, watching in silence as she wrapped its body in a beach towel. They crowded around her as she loaded it into the back seat of her Toyota. As the car pulled out of the driveway, they followed, dragging their feet on the asphalt, like when a loved one passes away and you trail along behind as they leave the house for the last time, not wanting their body to go on alone.

NOTES

1   *Mahalo* to cultural practitioners Kealoha Pisciotta and Roxanne Stewart for the title of this vignette, *Kai Palaoa,* a sea of whales. They use the word *palaoa* to refer to all species of whales. Their work honors the significant role of whales in Hawaiian creation within the Kumulipo, as manifestations of the ocean God, Kanaloa, and as ancestors. The appearance of a school of whales, *kai palaoa,* a sea teaming with whales, is an extraordinary event.

2   The first historical sighting of melon-headed whales in Hawaiʻi was in Hilo Bay in 1841 (Brownell et al. 2009). Mainly these whales stay in deeper waters, and they have only occasionally been sighted near shore in the Pacific, sometimes before mass stranding events.

3   Studies later attributed the stranding to navy use of mid-range sonar in waters surrounding Kauaʻi (Brownell et al. 2009). On October 13, 2017, a school of pilot whales was spotted near Kalapaki Bay on the south shore of Kauaʻi. Although some of the whales that swam close to shore were shepherded back into deeper water, others stranded themselves on the beach and the rocks of the breakwater. At least five *palaoa* died (Buley 2017).

CHAPTER 3

Kahu

Care and Cultivation

[Our family] had their own taro and they lived close to the beach 'cause
you could go fishing, you could go get 'opihi, everything was here.
—Loke Pereira, Anahola, 2007

We normally used to fish mostly down in this area and leave that
[nearby reef] for them.
—Tommy Hashimoto, Hā'ena, 2010

You learned the laws of nature, of when to harvest. [You] need to
know the cycles of whatever you're harvesting.
—David Sproat, Kalihiwai, 2015

Take care, before you take.
—Noah Ka'aumoana, Hā'ena, 2016

The word *kahu* means "keeper," and refers to the action of caring and nurturing. This term is commonly used to describe the individual or family
who cares for a particular *'āina*, especially sacred areas throughout Hawai'i.
Kuleana associated with fishing includes the responsibility to *mālama* (take
care of) fishing spots before harvesting from them. Use of natural resources
relies on building relationships with specific harvesting areas and gathering
in a way that cultivates their continued abundance. Just as North American
temperate forests once thought to be wild were actually tended by Native
peoples for generations,[1] small reef and inshore areas of Ko'olau and Halele'a
have long been tended by area families.[2]

These families fished and gathered mainly from small patches of reef
within a larger *ahupua'a*. Sustained interactions with these reefs allowed
fishers to develop in-depth knowledge of these spots and to observe their
changes across generations. Families share stories of protecting fishing areas as hatcheries and feeding grounds, timing harvests to protect reproductive periods, and weeding and cultivating seaweed beds as the base of the

marine food chain. People respected other families' gathering areas, leaving fishing spots to be harvested primarily by their caretakers. Aspects of this system and the underlying values and caretaking on which it is founded continue to be practiced today.[3]

Kuleana (Rights and Responsibilities)

Exclusive *Ahupua'a* Fishing Rights

The first written laws of the Hawaiian Kingdom protected the exclusive fishing rights of local residents. *Maka'āinana* (*ahupua'a* residents) and *konohiki* (local-level chiefs) had exclusive authority to harvest and regulate fishing within *ahupua'a* fisheries, which stretched from the shore at low tide to the edge of the fringing reef, or a mile out to sea where there was no reef.[4] This system of local-level harvest rights and caretaking responsibilities was formally targeted when Hawai'i was annexed as a territory of the United States through the Organic Act of 1900.[5] With the intention of opening all *ahupua'a* fisheries for public access, the act required *konohiki* to formally register their "vested rights," lest they be lost. Though only one hundred of the three hundred to four hundred known *konohiki* fisheries succeeded in registering, the territorial government systematically pursued termination of *konohiki* and *ahupua'a* residents' rights through condemnation of these remaining fisheries.[6] Arguments for condemnation focused on exclusivity and food security, particularly during World War II, when opening fishing throughout Hawai'i was promoted as a means to overcome food shortages. Arguments against termination focused on the conservation value of locally tended fisheries that could in turn replenish other fishing areas. A series of territorial, state, and even US Supreme Court cases considered the legality of closing local fisheries to public harvest. In 1904, in *United States vs. Damon*, US Supreme Court Justice Oliver Wendell Holmes argued that the government could not terminate local-level fishing rights within an *ahupua'a* on O'ahu.[7] He wrote,

> A right of this sort is somewhat different from those familiar to the common law, but it seems to be well known to Hawai'i. . . . The plaintiff's claim is not to be approached as if it were something anomalous or monstrous, difficult to conceive and more difficult to admit. . . . However anomalous it is, if it is sanctioned by legislation, if the statutes have erected it into a property right, property it will be, and there is nothing for the courts to do except to recognize it as a right.[8]

After Hawai'i became the fiftieth state of the United States in 1959, the state government continued to pursue payments to condemn remaining *konohiki* fisheries. Although Hawai'i's coastal fisheries are now managed as public resources, *konohiki* fishing rights are still legally recognized and reaffirmed in the Hawai'i State Constitution.[9] Some of the last registered *konohiki* fisheries in Hawai'i exist within Halele'a.

Formal protections for *ahupua'a* fishing rights were systematically eroded beginning in 1900, but respect for these exclusive rights of *ahupua'a* residents endured in Halele'a for a century longer.[10] When I asked Ko'olau- and Halele'a-area elders about where they fished growing up, they made comments such as, "We stayed in our own area." When I asked what the fishing was like in other *ahupua'a* just a few miles from people's homes, the most common answer was, "I don't know, we never fished there." When asked why not, elders named the families of the other *ahupua'a*, saying, "That was their place." These responses illustrate awareness of boundaries or *palena* that differentiate informal yet respected gathering rights reserved for local families. One Hā'ena elder, Uncle Tommy Hashimoto, laughed that if he had fished in other people's areas, they would have scolded him, asking why he was there and if his people had depleted their own reef at home. With the exclusive right to harvest in one's home area came the expectation of taking care of it.

Respected Family Harvesting Areas

Within the legally protected boundaries of *konohiki* fisheries, which spanned the entire coast of an *ahupua'a*, were much smaller areas informally reserved for families to harvest. For example, in Hā'ena, community members describe certain reefs, each of which have names, as being reserved for certain families. Generally these areas were in front of the lands on which these families resided. When asked where he used to fish, one *kupuna*, Charlie Chu of Hā'ena, stated, "Oh just in front of our land, on that reef there."[11] Ten years later, in my discussions with Chu's sixty-year-old nephew, Presley Wann, and thirty-five-year-old grandniece, Lei Wann, both echoed their uncle. Father and daughter pointed to the same reef Chu mentioned, saying, "Oh, I can only talk about this place,"[12] and "I grew up fishing right here and it is the only place I go."[13] While *maka'āinana* had formal rights to access all of the resources within their *ahupua'a*, most of their daily meals came from small family gathering areas close to home.

In some areas, gathering was restricted not to certain families but to certain groups of people. For example, one of the best and most accessible reefs for gathering seaweed in Hā'ena was reserved for *kūpuna* (elders). As *lawai'a* Uncle Bobo Ham Young explained, "That [reef] is considered the *kūpuna* place [to pick *limu*] because it's not rough. But you see these people [not from Hā'ena] go over there and they raid that place because it's so easy."[14] Other Hā'ena community members pointed out that while area fishermen consider it their responsibility to share their harvests with elders, many elders would still rather harvest for themselves. Respect for informal community agreements to save protected and accessible areas for *kūpuna* shows esteem for elders while preserving their independence. Reserving nearshore fishing for the people of a particular *ahupua'a* is another form of respect, preserving each community's ability to feed itself.

Access by Relationship

Family harvest areas did sometimes extend beyond *ahupua'a* boundaries in unique circumstances. Some species, for example, could be found only in one specific place, in which case provisions were made to allow people from outside the *ahupua'a* access to harvest. In some instances people harvested in more than one *ahupua'a* where they had family ties, often in adjoining *ahupua'a* along the same stretch of coast. People could also harvest in *ahupua'a* their families did not come from by seeking permission from certain respected individuals. For example, one *lawai'a*, whose grandfather, Kalani Tai Hook, was the *konohiki* of Wainiha, recalled people stopping at their home to ask his grandfather's permission before fishing in Wainiha.[15]

Individuals also fished with friends who had family ties in different areas. Uncle Valentine Ako grew up in a fishing family on a different island, but married a woman from Ko'olau and moved to Kaua'i nearly seventy years ago. He described learning to fish in one particular Halele'a *ahupua'a* from his friend who descended from that area. "He taught me all the *ku'una*s [fishing spots] down at [that place]. But you know, the way I was brought up, our *kūpuna* always told us when you go in a strange place, do not intrude. Go by invitation."[16] After his friend passed away, Uncle Val stopped accessing the areas his friend had shared, and refused to share them with others: "[My friend's] brother told me, 'eh, you ought to show your *mo'opuna*' [grandchildren]. But I said, . . . you know, the way I was brought up is never to intrude in a certain *ahupua'a* for the people who live there. And that's the reason that I didn't share. . . . I was brought up a different way, to respect the places that you associate."[17]

This fisherman felt that sharing knowledge of a place he wasn't from would show disrespect for that place and its people. He held to this value

Uncle Valentine Ako with the author and her children at an anniversary *lūʻau* for
Uncle Val and his wife, Aunty Elizabeth Ako. (Photo by author)

even though a person of that place, the brother of the friend who generously taught him, was urging him to share those teachings. Families from
this area use the word "respect" frequently in describing interactions with
place. Here, respect means that if someone shares a secret family spot with
you, you do not go back without them, or show others. Honoring the rights
of area families was a valued form of respect. Aunty Annabelle Pa Kam
elaborates: "I'm from Kalihiwai. For me to come over here [another place]
to fish, I have to know somebody over here and have permission. You just
don't go help yourself, because that's for them [the people of that place].
Everything is respect."[18]

Today, hundreds of people come from outside of Haleleʻa every day to
fish all along the coast. However, many Haleleʻa *lawaiʻa*, including some in
their twenties and thirties, continue to fish, gather, and even go to the beach
only in the areas where they went with their families growing up. If they
have a specific reason to go to another area to fish, they first seek permission
from the oldest area resident. Respecting family areas and restricting harvest to the places where your family comes from or has *kuleana*, the right
and responsibility to care for an area, continues to be an important value
underlying fishing.

'Ike 'Āina (Specific and In-Depth Knowledge of Place)

Exclusive harvesting rights meant that families harvested in small areas, sometimes referred to as a family's "icebox," and got to know these areas well. Lei Wann, a young fisherwoman of Hā'ena, explains how her grandmother would go to a particular spot, during the right season, at the exact time on a certain tide, to catch a particular fish. A Kalihiwai community member also recalled the specificity of fishing as practiced by her grandmother and older aunties:

> We would go to [the beach] . . . where I got my . . . reef fishing experiences. With the old grandmas, . . . high tide, we would all go. Put a kerchief on their face and their long sleeves. They didn't look like they were going fishing. It looked like they were going shopping. You know how the old ladies did, they would just cover up. . . . And then I watch these ladies in action. . . . And they would fish for specific fish. Not "We'll go and see what we are going to get." They would go for specific fish.[19]

Hā'ena *kupuna* and master fisherman Tommy Hashimoto says, "I don't look for fish. I meet them when they come home for lunch."[20] He goes on to explain that fish, like people, follow routines, set by tides. Certain species frequent specific reefs to feed on particular seaweeds. By knowing the depth of water on these reefs at certain tides, and how much water particular species need to feed, fishermen know when and where to find these species like clockwork.[21] Tommy Hashimoto created a map of over 160 Hā'ena reef names he learned from his father, though he says his father knew many more. This map, with names such as Kalua'āweoweo (hole of the *'āweoweo* fish) and Laeokahonu (headland of the turtle) highlights specific knowledge of multiple spots on each reef.

Kainani Kahaunaele describes knowledge that comes with harvesting from specific places across generations by saying, "I think the families down there were all like resource managers in a way. . . . They have been there for so long that they have seen the long-term changes. And because they are so observant, because they have to be in fishing, they will notice the nuances and so on."[22] Fishing provides the foundation for caretaking. Regular interaction with particular areas over long periods of time helps fisher men and women understand natural cycles and rhythms, recognize changes, and identify times to rest areas so that the resources may replenish. With exclusive harvest rights and in-depth knowledge comes the *kuleana* to ensure abundant harvests for months, years, and generations to come.

Māla ʻAi (Garden): Cultivating the Sea

Enhancing Habitat

Interviews reveal how families carefully tended nearshore coastal areas to cultivate abundance. One focus of this cultivation was creating and maintaining habitat for particular species. For example, building *imu kai* in the ocean in Hāʻena to create habitat.[23] *Imu kai*, also called *umu* in other parts of Hawaiʻi, are rocks arranged in small piles near shore.[24] Coral chunks with *hakahaka*, or large spaces, are most effective, as they provide numerous holes and hiding spaces for fish. *Kūpuna* of Hāʻena used *imu kai* to harvest, placing a net over the rock pile while removing each *pōhaku* (rock), in turn, rebuilding the pile stone by stone in another spot. When all the rocks were removed, the net was full of fish that were hiding in the original *imu kai* and a second was already built.[25] One of the last Hāʻena fishers who harvested in this way was Kupuna Rachel Mahuiki, who passed away in 1996 at the age of eighty-two. Kupuna Rachel, who gathered on the reef well into her seventies, would reach into the *imu kai* to grab *manini* (convict tang) with her bare hands, then place the fish in the folds of her *muʻumuʻu* (long dress) and walk home.[26]

In 2013, the Hāʻena community began rebuilding *imu kai* in an area known as Kalei.[27] Efforts by a summer student intern to experiment with various shapes and forms of *imu kai* (including tubes, pyramids, and cylinders) found that each shape attracted different species of fish. All shapes increased the abundance of multiple species in areas around the *imu kai*.[28] *Imu kai* built in nearshore waters aggregate and create safe habitat for fish species.

Another example of cultivating coastal areas is the maintenance of *heʻe* (octopus) habitat by women. Historically, mainly women harvested *heʻe*, an important source of protein that was pounded or boiled to tenderize the tough meat, then often dried. Women described how they would chew coconut then spit its oil on the water, making the surface glassy so they could search the reef for holes where octopus lived. Annabelle Pa Kam, whose grandmother taught her to be "her eyes" to spot *heʻe*, said that, with the coconut, "even if there is wind, everything is clear."[29] Her grandmother showed young Annabelle how to recognize octopus holes and clean them. If the holes were empty, granddaughter and grandmother would move the rocks to make it inviting for the octopus that lived there to return, or for a new one to come and make a home. Most area families reported harvesting octopus by stick or by hand, not by spear, tickling the octopus to grab hold so they could be lifted into a net or boat. One reason for not spearing an octopus was that, once it died in a hole, other octopus would never return to live there. Another reason to harvest by hand may have been to avoid spearing females sitting on eggs. By tickling them out of their holes first, fisherwomen could check for eggs instead of finding them once the mother had already been killed.

Ho'omalu: Protecting Reproductive Areas and Cycles

Protecting reproductive cycles was another key element of caretaking. Families described avoiding harvest of multiple species while they were spawning. "We would see the eggs, and then we'd stop, cause we know we [are] not going to have supply if we continually take them when they're *hāpai* [with child]."[30]

Places important to reproductive cycles were also protected. In Hā'ena, a lagoon called Mākua, meaning "parent," was recognized as a spawning and nursery habitat. Seventy-year-old Kapeka Chandler recalled her father teaching her to walk in the tree line at Mākua, rather than on the sand or along the shore. This way her footsteps would not scare baby fish from the shallows into deeper water that predators could reach.[31] However, since the 1970s, Mākua has become a high-traffic visitor recreational area. During the 1980s and early 1990s, twenty zodiacs per day launched through the reef to ferry visitors down the Nā Pali Coast. Hundreds of tourists snorkel and swim here daily. Hā'ena fisherman Jeff Chandler explained the impact of recreational uses at Mākua this way:

> [It takes only] ten seconds, a shadow, for the fish to react. One second you are in there, the fish are all out of there. They don't just come back. [They are] wild creatures.... I surf too, and I know I have an impact as a surfer. I know surfing impacts them. It takes two to four hours before fish come back. More impact from everyone, less chance they go back in.[32]

Though surfing is a traditional Hawaiian sport, like many aspects of the culture, it had nearly disappeared by the 1950s. Area residents recount that when Californians began to show up with surfboards to try out north shore waves in the 1960s and 1970s, elder Hawaiian fishermen worried the sport would negatively affect fish populations. One *konohiki* chased away surfers, including one who became his own son-in-law, to keep them from disrupting spawning patterns in an area bay.[33] Keli'i Alapa'i, a Hā'ena fisherman, remembers being excited to try surfing as a teenager. "We had to hide our boards in the shrubs on the beach whenever one of the older fishermen came by. If my uncles saw me with a surfboard I would catch it."[34] The ocean was viewed as a place for harvesting and cultivating food, not for recreation. As multiple community members put it, "Would you play around in your icebox? No you would mess up your food." Natural cycles such as spawning were to be interfered with as little as possible.

However, some species, including schooling fish such as *akule* (bigeye scad), could be harvested only at spawning times because spawning brought them into sandy bays within reach of fishers' rowboats. In one particular bay within Halele'a, a *konohiki* family regularly oversaw thousand-pound commercial *akule* surrounds. These fish were packed on ice, driven to the airport, and flown to Chinatown in Honolulu well into the 1960s, providing a steady source of fish for O'ahu's urban population. How were such large harvests sustained year after year without depleting the resource? Family members credit in-depth *konohiki* knowledge of *akule* spawning and aggregation behavior. The nephew of the *konohiki* described his uncle directing harvests from the cliff where he could see the entire school. This nephew and other younger family members waited on the rowboat, watching the *konohiki*'s signals to direct when and where to lay the nets. From up on the cliff, his uncle surveyed the colors of the school. The school normally appears almost black in the water. When the females start laying their eggs at the bottom, they stir up the sand with their tails, lightening the color of the school to brown. The color turns red as the males fertilize the eggs. The *konohiki* waited till the color of the school faded back to black, signaling that the reproductive cycle was complete. His nephew recalled, "When [the *akule*] had done all their spawning, that's when he would give us the go."[35] When *konohiki* rights were condemned and anyone could legally surround *akule* in the bay, local fishermen began to hurry to surround the schools before others caught them, without giving the *akule* time to reproduce. As a result, community members say that schools today are nowhere near the size of those in the bay fewer than forty years ago. Before, it was common to surround a thousand pounds at one time and still have taken only a small portion of the school.[36] By allowing the *akule* to spawn before harvesting them, this particular *konohiki* family sustained large harvests throughout the 1970s. Area residents recall that even with these huge harvests, it was common to see a sizable school of *akule* in the bay on any given summer day.

Hanai I'a (Feeding Fish)

Another practice was to feed fish in certain areas. These feeding spots were often called *ko'a*. Interviewees on Kaua'i used the word *ko'a* to describe areas where certain species gather. *Ko'a* were used extensively in places such

as Miloliʻi on Hawaiʻi Island, to feed ʻōpelu (mackerel scad) and other species.[37] As in Miloliʻi, some fishers on Kauaʻi tended these areas by feeding
fish. Aunty Linda Akana Sproat of Kalihiwai recalled a Filipino fisherman
named Levy who lived near Hanapai Channel in the 1940s and 1950s. He
would walk around the rocks from Hanapai to Aunty Lindaʻs familyʻs home
and ask her grandmother what kind of fish she wanted for dinner. Then he
would paddle his small canoe out to the bay. Inside, he kept a set of wooden
mallets of varying sizes. Each mallet made a different sound underwater
when tapped against the side of his boat. By releasing food into the water while tapping different mallets, he trained different species to respond
to each tone. When he tapped a certain mallet against the side, the corresponding fish swam right to his boat. He always brought whatever species
her grandmother requested in plenty of time to prepare the fish for dinner.[38]
While this was the only instance shared of this specific practice, similar stories of fish feeding are described across Hawaiʻi, and may have been practiced by other Haleleʻa families as well.

Tending *Limu* Beds

Limu (seaweed) is important in the diets of Hawaiian families, providing
a vital source of iron and other nutrients. *Limu* also provides the primary
food source for herbivorous fish, and many fishers describe seaweed as the
foundation of a healthy marine ecosystem. Haleleʻa and particularly Koʻolau
were known for their abundance of *limu*. Many elders describe having
abundant and diverse *limu* beds well into the 1980s, including *limu* species dependent on freshwater infusions into the ocean, which are scarce
or no longer occur today. Land-based sources of pollution—including fertilizers and herbicides from the sugar plantation, lawns, and one area golf
course—are cited as the cause of this loss of *limu*, along with sediment from
construction activities, spread of invasive species, irresponsible harvest
practices, and decrease in stream and groundwater flow caused by stream
diversions and private wells.[39]

Harvesting practices are specific to area and type of *limu*. These practices
include conservational measures such as rubbing seaweed against pant legs
or another surface after gathering to release spores back into the ocean, or
picking rather than uprooting the *limu*, leaving the holdfast[40] to regenerate.
Though specific practices for harvesting different species varied in different
places, Aunty Annabelle Pa Kam captures the general deliberateness and
care required: "If you pull the *limu* rough, come *kāpulu* [careless, unclean].
If you are greedy you only get ʻōpala [rubbish]. [Thatʻs why] today no more

limu. [People these days] don't care how they pull it. You pull easy, by the edge like this, come out easy."[41]

Abundance of seaweed was also cultivated by weeding certain reef areas, as with gardens on land. Aunty Jenny Loke Pereira's father, Andrew Lovell, held community-recognized *konohiki* fishing rights for different species of fish in multiple *ahupuaʻa* through the 1960s. She does not recall him regulating harvest of *limu* in these areas or requiring people to obtain his permission to gather. He did require his daughters to weed *limu kohu* (an especially prized seaweed) beds of at least one area bay just before the growing season, "Our days, during high school we used to go in the month of May before summer to start cleaning the reef, all the *ʻōpala* (rubbish), we used to pull it all out."[42] She felt that this work contributed to abundant populations and harvest by removing unhealthy patches, strange *limu*, and *ʻōpala* at a crucial time in the species' growth cycle.

Aunty Loke Lovell Pereira with her husband, net-maker Charlie Pereira. (Photo taken at their great-grandson's first birthday *lūʻau*, courtesy of their granddaughter Kanani Velasco)

For *limu kohu*, large harvests occurred in the summertime when the species was nearing the end of its growing season, before long days and intense sun dried out and killed the *limu* beds. Many *kūpuna* from this part of Kaua'i recall spending high school summers in the 1940s and 1950s gathering *limu kohu*, filling up metal cracker tins for sale in Honolulu's Chinatown. Some *kūpuna* used this money to pay for school on O'ahu. Those returning to school carried their harvest with them to sell in the city, then sent what money they could spare home to Kaua'i for their families. Like *ahupua'a* fisheries, *limu* harvests were an important source of both subsistence and commercial sustenance for area families. Large harvests were sustained through caretaking with specific management practices such as weeding *limu* beds early in the growing cycle and concentrating harvest at the end.

Rotating Harvest Areas

Like farmers who allow particular fields to lie fallow, Halele'a families also followed informal, self-enforced systems of rotating harvest. Families frequently let fishing spots rest after harvesting them to allow fish stocks to recover. As one fisherman, Bobo Ham Young, whose grandfather Kalani Tai Hook was the *konohiki* of Wainiha, explained, "That is how the old folks did it. Grandpa did not fish certain places and he told all the uncles, 'Don't go fish over here for certain months out of the year.' And sure enough, they don't fish, and when they go back, ah! The *i'a* [fish] stay home again."[43] When asked how to restore the fishery, elders suggested restoring seasonal rotations, "Make sure they *kapu* [protect], a certain season. Give the fish a chance to come back again."[44]

Lawa Pono: Take Only What You Need

> It's about taking what you need, never pillaging the spot, because once you take a resource and it's gone, it's gone forever.
> —Makana Martin, twenty-year-old Hā'ena fisherman, 2009

Lawa pono means "enough." The value of "take only what you need" was a key principle underpinning *kuleana* for taking care of fishing areas. Cultural expectations to cultivate restraint in harvest were mentioned by nearly every elder I interviewed, as well as by younger generations of Halele'a fishermen and women. When asked in a 2009 meeting to write down traditional Hawaiian values that guide fishing, fourteen out of sixteen Hā'ena area community members wrote, "Take only what you need." As Jerry Kaialoa Jr.

described, referring to the teachings of his well-known fisherman father, "You got to fish only what you can eat—that is what you bring home."[45] Harvesting just enough for a family's daily meals was a necessity before refrigeration. As Annabelle Pa Kam explained, "We ate fish, crab, 'opihi [limpet], squid, everything from the ocean. We got things fresh. We only took enough to eat. Never had refrigeration. Never had electricity. So we only caught what we could eat for the day."[46] Many interviewees felt the availability of freezers coincided with reductions in the health of area fisheries because they allowed people to overharvest. Still, interviewees expressed the importance of cultivating restraint. "Don't go fish every day. What you have, last you a week or maybe days. Then you can go back. But you take care of this one first. No go and throw 'em in the freezer and forget about this."[47]

Customary management in Halele'a did not define specific catch limits, relying instead on broad norms such as "don't waste" or "take only what you can use," which depended, for example, on family size.[48] If people did have more catch than their family could eat, they shared. In communal surround net harvests, which yield meals for many families plus a surplus to dry or freeze, fishers could employ catch and release to avoid taking the whole school, though today, many do not. One man described his father-in-law, a master fisherman of the Hā'ena area, ordering the release of part of a school of fish in a 2008 surround net harvest:

> We took what we could take first. Then when we brought it into the shoreline, we look and Uncle said, "That's enough." And maybe it was only half of the pile and he said, "Just let them go." We didn't scare them or anything so they just stayed in the outside net . . . acts like a coral. . . . So we go out there. . . . we just pick up the net and we scare them [the fish] and they going run home.[49]

Whatever species were caught were utilized in their entirety. For example, multiple Wanini families shared that if they harvested a *honu* (turtle), they used every part, including eating the meat, using the oil as a medicine to treat burns, and saving the shell for diverse uses, including bathing a child, until harvest of these turtles became illegal under the Endangered Species Act of 1973.

Many community members said that specific catch limits and rules were not necessary when they were growing up between the 1920s and the 1970s: "Never had rules, had so much, didn't know about this kind of stuff. Just took what we needed."[50] Some said that if they are not greedy, the ocean will always provide: "When I go fishing, I'm not thinking about how much I need to catch. I just say a *pule* [prayer] and throw my net. Whatever I get is exactly what I need. It is always enough for the family or party I went to catch for."[51] These interviews describe a broad underlying value of restraint in harvest that, when adhered to by a small population of users with exclusive harvest rights, left little need for formal regulations.

This general principle of minimizing impact, consuming no more than the minimum required, extended beyond fishing to all aspects of life within the community. Living within limits was a key part of good caretaking and ensuring sustenance for future generations.

> You don't need to take more than what you need for that day. You don't need to turn it into some kind of profit that is just going to benefit you and nobody else. So take what you need, just look at your daily and most immediate requirements and satisfy that. Because if you do it that way, in conjunction with a proper management system, then you're always going to have enough. . . . The *kūpuna* definitely thought in a multigenerational perspective, and that's why they only took what they needed.[52]

Changing Values

> *In our days when you see any ʻōpala, you pick it up yeah? You pick it up and put it on the side, you put it somewhere where it belongs, but you don't just leave it. . . . This generation is altogether different. People ignore what they needed to protect.*
> —Loke Pereira, 2007

Some *kūpuna* I interviewed expressed concern for the actions of younger generations. Nowadays, many younger fishermen, along with newer residents who have moved to Haleleʻa and Koʻolau, fish larger areas, targeting places known to have abundant fish, rather than sticking to certain spots. The value of earning the right to fish in a place through caretaking is less practiced and increasingly less understood. Values that apply more generally to behavior in *all* places, such as *mālama ʻāina* (take care of the land) are important, but lack relationship and reciprocity with a specific place. The general principle of not taking too much is widely talked about. However, as one Hāʻena fisherman in his thirties, Atta Forrest, asked, "We always say 'Take only what you need.' What if 'what we need' is now too much?" Knowing the boundaries of sustainable harvest requires getting to know and live within the limits of specific spots.

Haʻawina (Lessons)

The stories shared in this chapter teach that *kuleana* extends to caring for the natural resources that sustain you. In Haleleʻa and Koʻolau, *kuleana* to care for family fishing areas was exercised through providing habitat for marine species, by weeding and cleaning the reef, staying within the boundaries of one's own cultivated area, and limiting, rotating, and timing harvests according to reproductive cycles to ensure that resources were able to regenerate. People respected the work, knowledge, and gathering rights of *kahu*, families who invested time in tending their own areas of harvest.

What would it look like in today's world if every family took responsibility to actively care for the places that nourish them, whether these places provide food or wonderful memories together? What if everyone was a keeper of some place, tending its soil or shoreline in a daily, hands-on, and intimate way? What if our worth was assessed by the health of the natural resources in our home areas? When we are reliant on our relationships with a particular place, we are also more likely to regulate ourselves, to refrain from eating a certain species when it needs to be rested, and to live within limits.

Kuleana is more than a general ethic of care for the Earth, minimizing your impact wherever you go. It encompasses distinct responsibilities to the specific places that nourish our lives and families and the right to *mālama*, or care for, these places.[53] *Kuleana* grows from reciprocity: regular return, cultivation of relationship, and active work to nurture abundance. Caretaking renders certain places "ours," while making us more conscious of our behavior when we are in places cared for by others. Which places are you and your family responsible for? Will they continue to flourish, to nourish you and allow you to feed others?

Moʻolelo: Lawaiʻa (Fisherman)

Luʻuluʻu Hanalei i ka ua loku, kaumaha i ka noe o Alakaʻi
Hanalei is drenched in the pounding *ua loku* rain, weighted with
the mists of Alakaʻi (an expression of heavy sorrow)

One summer while home from college, I ran a cultural camp for Hawaiian youth
and lived at Waipā behind the poi garage. The shack I lived in was falling apart,
and a bit overwhelming, with many visiting rats. But Uncle Calvin Saffery always
checked in on me, offering food and making sure I was all right. At the time, Uncle
Cal lived at Waipā too, in front of the warehouse. For him, this was a return to
Waipā, where he was born. In return for his living space, he cooked the taro roots
delivered by area farmers each week before poi day. Every Thursday at Waipā, community volunteers assemble to clean the taro to make poi, a Hawaiian staple food.
Waipā staff sell the poi at low cost and deliver it free to *kūpuna* throughout Kauaʻi
to enhance community health and access to traditional foods. I loved listening to
Uncle Calvin's stories while he worked, building the fire in the pit under the metal
barrels full of taro, splitting wood with sure strong swings of the ax and patching
his fishing nets.

Uncle Calvin Saffery could feel the fish coming, long before anyone else
saw them. I used to watch him at night, meticulously mending his nets
with the bamboo needle. Occasionally, he took a deeper breath of the
night air as if it held clues to where the fish would be in the morning. I
marveled at his gift.[1]

I remember him racing home to Waipā from his job cleaning our three
county parks, raking up every leaf from the false *kamani*, because he felt the
fish. Uncle Cal would switch from his rusting green county truck to his even
rustier blue clunker truck and spin off. He'd return three hours later with
his catch. If he passed me in the yard at Waipā when he drove in, Uncle Cal
would lean jauntily out the window on his elbow, trailing one word—the
name of the fish he just caught, "*ōpelu*," "*akule*," or "*ʻanae*" (mullet).

Everyone in Hanalei seemed to know when uncle had gone fishing, and
cars would be waiting in his driveway by the time he pulled in. People knew,
too, that the only price for Uncle Cal's fish was listening to the legend of
how he caught them, in which the hero was always the fish. He'd hang off

the back of the truck giving *iʻa* (fish) away by the cooler, his warm voice booming, "These buggers better be *ʻono* [delicious] because they sure made me work this time, these are some smart *ʻōpelu* you'll be eating." And, "Make sure you take for your mama too."

The blue clunker truck, so rusty now it does not move from its space at Black Pot Park, has become Uncle Cal's home. The *manō* (shark) tattoo of his *ʻaumakua*, striking from his bicep and shoulder to his chest, is becoming flaccid, as his skin caves into his bones. He had to be evicted from his house at Waipā and left his nets. A few of us who worked there spent two days folding, cleaning, and wrapping the nets into tarps. We stored them out of the rain to keep safe for him, though I don't think he realizes that the nets are gone. His breath is raspy now with a cough that never goes away and his eyes slide quickly in many directions behind his sunglasses.

Still, he is always happy to see me, and on Saturday, when I locked my keys in the car at the beach park, he helped me break in with his characteristic jaunty gallantness. Once on the road again, I found myself in Big Save wanting to thank him. I bought *poke*. It wasn't until I got back to his truck that I worried if it was offensive to bring fish to a fisherman. I was spared finding out, because when I returned, Uncle Cal was fast asleep, one booted foot sticking out the driver's side window. I looked in the back for a cooler and found one with ice that had long since melted into rancid slime with dead mosquitoes floating in it. The smell of fish clung to its weathered plastic.

On the day of Uncle Calvin's funeral at the green Waiʻoli Huiʻia church in Hanalei, a slideshow runs pictures while a line of well-wishers waits to hug his sisters and grandchildren. In the photos, Uncle Cal is strong and larger than life. He leans over a tiger shark he caught, carries huge mountain pigs on his back after a hunt, and digs *imu* (underground ovens). He holds a hammerhead in a headlock at the mouth of Hanalei River and drives a boat in cowboy boots. In one shot, Uncle Cal stands in his wetsuit with a string of huge *kala* (unicorn fish), a cigarette hanging from the corner of his mouth.

There are pictures of his later days too, the more recent years, when he recovered from his drinking and other addictions. Then, he got a Hawaiian Homes lot in Anahola and built himself a house, board by salvaged board. He is thin in these photos, a man who had wasted away but fought his way back. There are photos of him surrounded by grandchildren. In a few pictures, his four-year-old grandson sits on his lap. They rake leaves together and play ukulele. He looks adoringly at the children in these photos, and I can almost hear his teasing voice and raspy, contagious laugh.

There are many empty seats in the church pews, not because few people have come, but because most are more comfortable outside. Hundreds stand on the grass, talking in clumps, preparing trays of food, setting up benches and tables—jobs Uncle Cal would be doing too. The young men of the north shore are here in force, every single one he taught to fish.

Uncle Calvin was a hero and role model to many. He knew this whole coast like the back of his hand. He had lots to teach and little respect for rules. When schools of *ʻoama* (baby goat fish) came in to the shallows of Hanalei Bay in the summertime, many eyes tracked where the pile was. Bamboo poles showed up along the shore and lined the pier as people patiently hooked fish one at a time, the only legal harvest method. Uncle Cal had no patience for hooking fish. He was known to go at night and throw his net on the whole pile, after which every community event for the week, including poi day at Waipā, would have a full pan of fried *ʻoama*, *kūpuna* relishing them like chips.

At Uncle Cal's funeral, the tables in the mission hall next door to the church are loaded with food. There is a whole table of dessert. Another table is all fish: *kala* soup, fried *akule*, *nenue poke*, *lomi ʻōʻio*, platters on platters of sashimi, a tray of *ʻoama*, and the crack seed *palu* of fish entrails his mother, Aunty Jenny, was famous for.[2]

I remembered the days Uncle Calvin came home from to Waipā from a *hukilau*, coolers brimming with *ʻōpelu*. I used to dream of tracing the path of any one fish from the plate, back through all the layers of giving, to Uncle Cal. His was the kind of generosity that allowed everyone he gave to, to be generous in turn, so that a whole community together, even if at different tables, would be eating *ʻōpelu* for dinner.

NOTES

1   Uncle Calvin's father, Jack, was a cowboy for the Robinson family, who leased Waipā from Kamehameha Schools Bishop Estate to run cattle. His mother, Aunty Jenny, once told me that the first (and possibly only) time she ever drove a car was when she had to pack all her kids into their Rambler and drive to high ground at the cemetery in Waiʻoli before the tidal wave hit in 1957.

2   *Poke* is a Hawaiian dish made with cubed, raw fish mixed with seasoning—typically soy sauce, sesame oil, green onion, white onion, *kukui* relish, and *limu*. Simply fried with salt is a locally favored way of eating fish, as well. *Palu* is chopped-up fish, primarily *aku* (bonito), flavored with the eyeballs, bloodline, and innards. "Crack seed" is a Chinese-influenced version of *palu* made mainly in Hanalei with *akule* or *ōpelu*. In crack seed, the bones and meat of the fish are chopped up and flavored with garlic, shoyu (soy sauce), and sometimes chili pepper.

I have mainly eaten "crack seed" prepared by Aunty Jenny Saffery, mother of both Calvin and Kealoha, featured prominently in the next chapter. Aunty Jenny once told me that she never had the chance to go to school but her "smarts" were in her tongue. She could go to any *lūʻau*, taste any dish, and go home and re-create it. She was known for her crack seed, her cakes, including pistachio, and her mischievous sense of humor.

Aunty Jenny Saffery, mother of Calvin and Kealoha Saffery, celebrating her birthday at Waipā poi day outside her family home. (Photo by author, taken around 2003)

Mo'olelo: 'Ōulilani, A Beloved Elder

This poem is written with aloha for Aunty Annabelle 'Ōulilani Pa Kam, beloved rascal, regulator *kupuna* of Kalihiwai, who passed away on October 12, 2015, at the age of eighty-two. I was blessed to interview her multiple times over the past fifteen years. She was always willing to talk story and generous in her sharing. This poem brings together language from transcriptions of those interviews, mainly her words. I wrote this poem for her memorial service, when her family asked me to speak. Each section starts with one part of the many meanings of her Hawaiian name, 'Ōulilani, and shares some of the lessons she shared with me.

'Ōulilani: Tūtū

Your Tūtū called you 'Ōulilani
Woke you gently ahead of the dawn
To cross the river, when the reef was dry
To be her eye for *he'e.*
Chewed up coconut spit on the water
You can see everything, even in wind
The reef was many colors then.
Tickle the *he'e*, no spear,
Clean the hole, inviting another to move in.
Anything for you to learn
She would call attention to it
You helped her gather *lauhala*
And drag the leaves all the way home.

Annabelle 'Ōulilani Pa Kam,
July 4, 1933–October 12, 2015.
(Photo by Mauliola Cook and
Sylvia Partridge)

'Ai lani: Spiritual food

You said, "Fish crab *'opihi* squid
Everything straight from the ocean
We got things fresh
That was our icebox
We only took enough to eat."

ʻŌʻuli: Character

> You said, "Our people
> Always kept everything clean
> The yard so big, my brother Chauncey on one side,
> Me on the other, pushing the lawnmower
> We had to get up early to feed the chickens
> Help Tūtū scrub the clothes on the board."

> "You know, you gotta be rich to own toys
> We had tin can play dishes
> Sardine tin boats
> Floating the river on strings
> *Ti* leaf sliding down the mountain when rained
> Kids don't do that kind stuff these days."

> "We never had to come home from playing
> If you hungry, crack *kamani* nuts
> Grab *hāʻukeʻukē* [urchins] off the rocks."

Kalihiwai, as Aunty Annabelle describes it, in 1944, two years before the 1946 tidal wave. Note the size of the beach, the three stores and homes located along the road that was then the main highway. On the ocean side of the road are vegetable gardens, a small store, and one family's home in which most family members were swept away and lost. Taro fields, which at this time were planted in rice, line the upper bends of the river valley. Hihimanu (the smaller peak) and, in the clouds, Waiʻaleʻale, Kauaʻi's highest mountain, tower behind. (Photo courtesy Bishop Museum archives)

"No money for the Chinese stores
Unless somebody came to visit us
When they did, we bought one lollipop
With our *kenikeni* [loose change],
And shared,
Your turn, now my turn
Five different kids taking licks
How wonderful it was
The way we were brought up."

'Ō'uli: Sign, omen

Brushing your teeth one morning
You watched the ocean rise up.
"Run," said Tūtū, "RUN!"
From the hill behind your house
You watched the water recede,
Then return high as the telephone pole
Houses swept away,
An entire family sleeping in one
Their father coming home from work,
To look for them.

When asked how you recovered,
You said, "We didn't *need* anything
We knew how to live simple
How to survive
That's what I teach my children, my grandchildren
How to live off the land."

'Ō'uli: Nature

You said, "Don't pull the *limu* rough
That's why today no more *limu*
You pull easy, by the edge like this
Come out easy."

"I'm from Kalihiwai
For you to come here to fish
You have to know somebody,
Have permission
You just don't go help yourself.
Everything is respect."

ʻAi lani: Majesty

> "I have a bad habit
> I'm very outspoken
> I say what comes to my mind
> If you don't like that,
> You have a problem."

ʻAi lani: To treat as a chief

> Tūtū said, "If you have something nice, share it
> To be rich is to give things away."
> Your house is the place,
> Where everybody came.
> You invited all of our friends over
> And fed them.
> Showed them your shell collection
> Your pictures, that view.
> Your boys made sure you always
> Had plenty fish to share.

Lani: Spiritual

> You said, "When Tūtū man died,
> Every Saturday
> We had to go clean grave.
> Tūtū would sit there and talk to him
> Then make us say something.
> 'Tūtū man take care us and make us be good,
> Because Tūtū said we naughty.'
> We thought, 'She's crazy, Tūtū crazy talk like that.'
> And today, I'm doing exactly the same thing."

> "When I brought my boys home
> From Honolulu, first thing
> We go introduce them to Tūtū.
> 'Here are your *moʻopuna*.'
> 'You tell her that you are not from here,
> You are coming to visit
> To see the place how I grew up.'"

> "Before we go back to Oʻahu,
> We gotta stop again and say '*Mahalo*.'
> Once a month,

Uncle Chauncey and I go grave together
I wonder who will go after we leave.
Tūtū you are 100 something years old
And we are still coming to talk to you
And ask you to take care of us."

"When you grow up with Hawaiian grandparents,
Everything is prayer.
You see how wise, so wise, yeah
When I look back now I say, So wise."

Ao uli: Sky blue vault of heaven

Since you left,
the Kalihiwai sky
has tried on every shade of blue
Ka'ōnohi sunset rays
stream through a *puka*, a space, in the clouds
skies matching indigo ocean
post sunset.
You said, "I could wake up here every day,
and it would never be the same."

A single *'iwa* [great frigatebird]
flies low over the bay,
rises up the cliff below your home,
skims toward Hanapai
where two other birds join,
rise, circle, soar
over trees, over ridge
through an opening
in the sky,
join a host of many *'iwa*
suspended, silhouetted
against sun lit
blue heavens

CHAPTER 4

Konohiki

Inviting Community Ability and Abundance

River house and ocean. That is how we was living. Get the river,
the land, and the ocean.
—Nancy Pi'ilani, Wainiha, 2007

Whenever I had a good catch we shared it with the neighbors.
Sharing was the thing to do with neighbors.
—Chauncey Pa, Kalihiwai, 2010[1]

You don't need to learn how to fish . . . to survive. That's why we have fishermen
and gardeners, taro farmers— because everybody got to work. . . . It's that bond of
sharing. I think that's what is lost—not lost, but just fading away!
—Kealoha Saffery, Hanalei, 2015

The word *konohiki* means to *kono* (invite, entice, or induce) *hiki* (ability), or to invite willingness.[2] In precontact Hawai'i, *konohiki* were local-level chiefs and chiefesses who—although varied in their rank and specific responsibilities—held *kuleana* for both physical and spiritual tasks to ensure the prosperity and abundance of an *ahupua'a*. While *konohiki* responsibilities historically spanned the entirety of an *ahupua'a*, from mountain to sea, references to *konohiki* in Hale'ea and Ko'olau from 1920 to 1975 referred mostly to fishermen with authority over certain coastal areas. Interviewees described these contemporary *konohiki* as regulating fishing, overseeing collective harvests, and sharing the catch. While the role of *konohiki* is often highlighted, these individuals were part of a larger network of community *kuleana* to share food, work, and talents. Community members' harvesting and sharing of resources across ethnic groups provided food, while also building strong relationships among families. Sharing generated collective abundance, local-level sufficiency, and resilience to recover from multiple natural disasters including tidal waves and hurricanes. Perpetuation of this cultural value of sharing to enhance community ability continues today.[3]

Changing Roles of *Konohiki*

Historically, *konohiki* worked with area experts (such as *kilo* or head fishermen, and master taro farmers) to facilitate decision-making about resource use.[4] *Konohiki* coordinated collective work projects, including maintenance of irrigation ditches for taro and annual collection of *ho'okupu* (offerings, literally meaning to cause growth).[5] Fulfilling the *kuleana* of a *konohiki* required many levels of expertise to invite willingness to share and collaborate with both the land and its people: "*Konohiki* had to know all of the *waiwai* [assets] contained within each *ahupua'a*—hydrologic, biologic, and geologic. They had to know the state of the soil, plants, and animals on land and sea, and guide decisions on their use. Most important, *konohiki* had to know how to deal with human beings."[6]

Coastal *kuleana* of *konohiki* were enumerated in Hawai'i's Declaration of Rights in 1839 and reinforced in subsequent kingdom law. Within *ahupua'a* fisheries, from shore to the coral reef or one mile out, *konohiki* could designate one species for their exclusive harvest or reserve one-third of all catch. In consultation with *maka'āinana*, they could restrict or prohibit fishing to allow the resources time to replenish.[7] Their actions were not to infringe on the ability of *maka'āinana* to feed themselves and their families.[8] By the 1940s and 1950s, contemporary *konohiki* of Ko'olau and Hale'a, some of the last registered and active *konohiki* fisheries in Hawai'i, exercised mainly only the first right, to exclusive harvest of one species per *ahupua'a*.[9] While much has been written on *konohiki* rights, interviews with family members of contemporary *konohiki* emphasize that these rights were rooted in weighty responsibilities.[10]

Hukilau (Collective Surround Net Harvests)

> *My dad, he spent a lot of time feeding everyone in this village. . . . My father was very generous. When he caught fish he always shared his catch with the village families and, you know, he supported a lot of them here.*
> —Loke Pereira, 2007, describing her father, Andrew Lovell, *konohiki* of Anahola, Moloa'a, and other surrounding areas[11]

> *[My father] gave the fish to the people first before he sell. He never keep the fish to sell first, no. The people come first. He was so kind, my papa. And the people, oh, they do anything for him. They love him because the way how he treat them. Not only one Hawaiian. Everyone from the end of the road to where we live. That's how my papa was.*
> —Nancy Pi'ilani, 2007, daughter of Kaho'ohanohano Pa, head fisherman of Hā'ena[12]

Koʻolau and Haleleʻa *konohiki* were responsible for ensuring the health of their fisheries, overseeing collective harvests, and facilitating sharing to provide food for the entire community. In the summers, *konohiki* and head fishermen routinely led five-hundred- to two-thousand-pound surround harvests (referred to locally as *hukilau*)[13] involving twenty to hundreds of people. Successful harvests required watching for schools entering the bays throughout the season, observing their behavior for days on end, determining when to surround, and mobilizing and directing labor, plus storing and maintaining fishing gear.[14] Before trailers, the families of head fishermen kept hundreds of feet of *hukilau* nets in boathouses near the beach and hauled thirty-foot rowboats loaded with nets across the sand to the ocean on wooden rollers. Surrounding a school could take weeks of watching the fish, and a full day of labor to deploy and haul in nets, remove the fish, and distribute the harvest.

In Kalihiwai, the *konohiki* fisherman "would sit up on the hill by the house and watch for the ball of *akule* (bigeye scad) to come in during the season. He'd just watch every day until the ball would come in—just this huge, huge massive ball of fish that would come and spawn in Kalihiwai Bay."[15] Though *konohiki* fishermen or *kilo iʻa* (head fishermen who spotted fish) directed the harvest, they could not surround fish without mobilizing the community. Aunty Bernie Mahuiki described how her father-in-law,

Aunty Linda Akana Sproat, whose father and grandfather were both *konohiki* of Kalihiwai, with her husband, Uncle David Sproat. (Photo courtesy of Raymond Kahaunaele)

La'amaikahiki Mahuiki, used his old policeman whistle to alert everyone in Hā'ena when a school came into Maniniholo Bay. He would scramble up the cliff on his trail behind the park where he was a caretaker: "If it was within reasonable reach for them to go and surround the fish, he'd blow the whistle. And you could hear that whistle from Hā'ena to Kepuhi. . . . And everybody that heard that whistle dropped everything that they were doing and they would say, '*Hukilau!*' We would drop everything and run down to the beach."[16]

Aunty Bernie says that "all of the families had a specific responsibility in the *hukilau*." A small group pushed the boat into the water, repositioning the rollers and two-by-fours repeatedly to reach the ocean. Oarsmen rowed the boat, while a netman watched the *kilo*. The *kilo* stood on the cliff waving white flags to signal when and where to release the net. As the oarsmen navigated the boat around the school and slowly moved it toward shore, divers made sure the nets were in place, kept them from getting caught in coral, and removed sharks or other unwanted catch. One community member describes helping with these different roles during a *hukilau* at Kalihiwai during the early 1970s:

> A few times I'd be in the fishing crew. We had these big wooden boats and we'd go out and [the *konohiki*] had thirty-feet-deep nets, and you'd row and row and row. The guy out the back would be throwing out the net. Then the boats would circle the ball of fish and the whole ball would be circled and it wouldn't realize it yet. Then all of a sudden they'd realize it and hell would break loose. Fish just jumping everywhere into the boat, hitting you in the head. And everything's in there. Whatever was chasing the fish, the shark that was chasing the fish, turtles, everything in there. And so you have to try to get out things like that. Then you'd bring the net closer and closer together, tighter and tighter. You dive down and sections of the net would come out and you'd hook the rest together. . . . You'd take out a section and get it tighter and tighter. By this time, almost the entire population of Kīlauea is on the beach, everybody—man, woman, and child—the entire population. And then the big bag net gets put in the water. This is a huge bag, maybe the size of a small house. And you'd go dive down and sew that into one of those joints in the net, then you'd pull the main net together until all the fish swam in this. And then you'd bring the boats together, the nets together with the huge bags of fish and somebody would swim in the big rope and then they'd yell, "*Huki* [pull]!"[17]

Once the fish were surrounded in the inner bag and the end of the rope was on shore, the large group on the beach would grab hold of the ropes and begin to pull in the nets. Aunty Bernie explains, "Yes, everybody knew

Hukilau, Hanalei. (Photo from Hawaiʻi State Archives, circa 1925)

Hukilau, Kalihiwai. (Photo by author, April 2015)

exactly what their responsibilities were. Ours was just getting over there and holding the net and start pulling." Surround net fishing required not only the *konohiki* or head fishing family but multiple families fulfilling different roles in the harvest. At the end of a successful harvest, many hands were needed to pull in the heavy nets and remove hundreds to a few thousand of pounds of fish.

Māhele (Distribution and Sharing) of Catch

> *Part of their upbringing was to share with the community. They'd feed all the families.*
> —David Sproat, describing his in-laws, the Akanas, a *konohiki* family of Kalihiwai, 2015

> *There was no limit to the māhele because his idea was to share his fish with everybody. When everybody had their māhele, then whatever was left over is what he would sell. He never sold any fish until everybody had their māhele.*
> —Bernie Mahuiki, talking about her father-in-law, Laʻa Mahuiki, a head fishermen of Hāʻena[18]

Once the fish were caught, *konohiki* and head fishermen were responsible for distributing the catch to area families and everyone who helped in the harvest. By the 1930s, surround net fishing was a commercial venture for some *konohiki* families, who sold fish to support themselves. Asian fish sellers came to *hukilau* to buy fish to sell to sugar plantation workers on other parts of the island. Some *konohiki* shipped fish to Honolulu, where it was sold in Chinatown, providing a source of fish to families from rural areas that had moved to the city. At the same time, *māhele* of fish caught in these harvests, including *akule* (bigeye scad), *ʻōʻio* (bonefish), and *ʻōpelu* (mackerel scad), remained a key food source for local residents. Linda Akana Sproat, whose father and grandfather were both *konohiki* of Kalihiwai, explained, "The people who came to help us, . . . grandpa would give them a share, *māhele*. If you came to work, everyone would get fish."[19] One community member recalled multiple motivations for going to help with *hukilau* as a child:

> We'd all go and *kōkua* [help] so that we could get our *māhele*. And in those days, no plastic bag or any kind stuff, so you take your T-shirt like this and you put your [fish] in your T-shirt and you run home. . . . But in the meantime, before you get your *māhele*, you can go swim, right? So if your parents ask you "How come you're wet? Where did you go?" "Oh we went *hukilau*." And we have the evidence to show that we went. . . . We brought dinner home, and then everything is okay.[20]

Commercial sales introduced tension over how much fish was sold versus given away to helpers and community members. Different *konohiki* and head fishermen were considered more or less generous in their *māhele*. Kalani Tai Hook, *konohiki* of Wainiha, is widely remembered for his sharing. His grandson recalled families coming to participate in the *hukilau*, bringing large bushel-sized bamboo baskets. After removing the fish from the nets, the families gathered around his grandfather, who stood by the pile of fish. Grandpa Tai Hook would throw fish to each family, one at a time, around and around in a circle, until all of the fish were gone and every family's basket was full to overflowing.

Some community members described squirreling away fish at surrounds where *konohiki* were not very generous. Some buried fish in the cool sand as they took it out of the nets, then came back to dig it up later. Others described catching species reserved for the *konohiki*, asserting that no one should care as long as they were catching only for their families to eat, not selling. *Konohiki* who were perceived to be less generous found it gradually harder to recruit labor for their surrounds and to enforce their exclusive harvest rights for certain species.

These instances show how authority and respect were based on fulfillment of obligations to both resources and fellow community members. Nancy Piʻilani remembers worrying aloud that her father, Kahoʻohanohano Pa, was being too generous in distributing fish from a surround and getting scolded as a result:

"Oh, that's enough," he tell the Japanese [fisher] man. "'Nough. This go for the workers" [everyone who helped to pull in the nets]. And he get all the fish for everybody. And they get most, then us. I tell my papa, "How come everybody get more fish than us?" "*Kulikuli! Kulikuli 'oukou. 'A'ole wala'au me kela* [Be quiet. Don't talk like that]." He tell everything Hawaiian. No talk. Because we asking him if we get 'nough? [He] telling, "We have enough fish. *Lawa* [plenty]. No grumble." He was so soft, my papa. Oh, I cannot get over him. He really aloha, real big aloha for the people.[21]

In the days before refrigeration, families' *māhele* was cleaned, salted, and dried. In the summer, when sun was plentiful and species were schooling, drying was a key means of storing food for times of less abundance, such as the winter months when the ocean was rough. Uncle Kealoha Saffery described the importance of drying to store seasonal foods such as fish that school in the summer for times of less abundance: "Fish, cleaning fish, everything was dry. You have to know how to dry the fish. Meat, most times it was salt the meat. That's the thing of how to survive. Those things I know, still know how to do from my grandmother."[22] Aunty Linda Akana Sproat remembers long lines of sticks set in the sand of Kalihiwai Bay. The sticks supported lines strung through the gills of *akule*, allowing the fish to dry in the sun while the ocean breeze kept the flies away.

Kūle'a: Sharing Skills and Ability

Research often focuses on *konohiki* as specific individuals and families, but *konohiki* actually represented a larger system of sharing to enhance community ability. Just as in surround net harvests, every individual and *'ohana* in Hale'a and Ko'olau fishing communities had a responsibility to share the "catch" of their particular skills with others. One root word of *kuleana* (responsibility) is *kūle'a*, meaning successful or competent. "When you have a competency, you have an obligation to use it for the betterment of society."[23] In this sense, "*kuleana* becomes the act of demonstrating aloha in your particular and unique way,"[24] by sharing one's work and talents. The word *lawai'a* comes from *lawa* (enough) and *i'a* (fish), reinforcing the idea that *lawai'a* are responsible for ensuring enough fish for all.[25]

Perpetuation of this cultural value of sharing continues today, extending to all community members and providing benefits far beyond food. Aunty Verdelle Lum described sharing as a sign of respect for the *kama'āina* (people of the place) and the *'āina* from which resources were gathered: "Wherever we go, we shared what we had. That's how the Hawaiian custom was. You catch, you go through somebody's property, be sure you give

Aunty Verdelle Peters Lum of
Wanini talking with students.
(Photo by Ron Vave)

them some. My grandfather and grandmother taught us that. You never go
through somebody's property and don't give them fish."[26]

Some describe the belief that sharing actually enhances fishing luck, saying, "The more you share, the more you catch."[27] Others describe sharing as
part of an underlying philosophy of giving:

> You need to give before you can take. And that philosophy in practice perme-
> ates the culture on every level. [Sharing] is just a further act of giving before
> you take. When we go fishing we take from the ocean but that's not just for us.
> We've got to give some more, give some more, give some more . . . all the way
> home. . . . You give to the ocean, you give everyone along the way until you get
> home and then, okay, there's something for us.[28]

The idea of giving more than you take, of having some skill or gift to share
with others, made for lives of resilience, surviving and thriving even through
lean times.

Lako: The Land and Sea Provide

> *My uncle had nothing, only taro, and they catch fish, and that's all their living was.*
> *Nothing else. . . . So you don't starve. Everything is there. Small little things, but you*
> *never starve. You don't have to go to the store. One time you go to the store, you*
> *don't have to go back for two months, maybe one month.*
> —Nancy Pi'ilani, 2007

Though many interviewees referred to their families as "poor," they also de-scribed lives of abundance thanks to their ability to obtain daily meals from the land and sea, and to share what they had. One Hā'ena elder, Aunty Violet Hashimoto, described Hā'ena in her childhood as *lako* (well supplied), its lands and reef providing everything community members needed. Another community member, Sam Meyer, recalled of his childhood in Wanini,

> Everybody that you talked to, the people that lived and grew up here, we were all poor people. And when people tell you that Anini has got to be one of the best places in God's Earth, I believe it, you know. . . . I mean if I had to do it all over again I would, in a heartbeat. You know, when you're going through it, you don't think of it as hard times. You don't think of it as anything. I mean you just survive, you live, you eat, you do.[29]

Interviewees in their sixties and older frequently told stories of going with their grandparents or parents to the ocean to gather for multiple meals a day. One woman described staying with her grandmother at Wanini: "We'd wake up early in the morning, because she went fishing every morning, to catch her breakfast." Her grandmother would go across the road from their home and catch a few of her favorite fish, *manini*. "Then she would go home and make her breakfast. She always had fish and poi."[30] Another woman in her eighties recalled her grandmother catching fish for the day's meals while out on the reef for a few hours gathering *limu*: "Yeah, that's where my grandmother used to come with her *'upena* [net]. And before she used to start to even pull *limu* she put the net right across the channel over there. So, when we pulling *limu* and the tide come up, the fish going out, yeah? And that's how she catch her dinner, her lunch and dinner" (Loke Pereira, 2007). Nancy Pi'ilani remembers full-day fishing expeditions down the coast with her father: "We never take enough food for us. But had water and what we did was go get the watercress, oh, so beautiful, green! Nobody had eaten you know, so at least we had something. And we'd go look for *'opihi* on the stone."[31] Uncle Kaipo Chandler recalled learning to eat *manini* as his father fished for them in Hā'ena:

My parents used to go fishing there, we would follow them. My dad would go net fishing and we'd . . . get scolded if we go in front, the fish all ran away, so we had to stay in the back and my mom always said that. And my dad kept a small bag *manini* where she eat 'em right behind him. . . . Raw. She'd shake them in salt water and . . . my mom would eat them and we eat too. . . . We follow, because my mom is eating. It looks so good so we try. Eh, we like it, so we all eat.[32]

Knowing how to feed oneself from the land was necessary to survive. The land provided plentifully, if simply. When families hit hard times, where they couldn't provide for themselves, other community members helped out: "Community, well, all local people. Japanese, Filipino, Hawaiian, few *haole* [Caucasian]. Everybody live together, they help each other. Never need to be told. They see with their eyes; they can see that person need help. They come, that closeness, everybody helps each other. They have their ups and downs but they were always there."[33]

To get others through hard times, in addition to direct food sharing, community members also shared survival skills with neighbors. Aunty Loke Pereira describes how her father always made a point of sharing fish with those in need—for example, widows. He also taught these women different types of fishing and gathering from the ocean that they could do themselves. Loke's cousin was widowed three times by "husbands that left early":

And she had to pick up the slack. She learned how to go pull *limu*, go fish and then just hustle, you know, and feed her family. She found a way to do that. And that's how they survived. 'Cause jobs wasn't that good those days you know? Either you went to work at the cannery or in the field, but it was not where you could really depend on. You had to go out and find food to feed your family, period. There's no turning back. . . . [These women] had to fend and they had so much spirit of surviving for their family.[34]

Learning skills to live off the land gave these women some measure of security to be able to provide for their families independently. Praising another elder woman, who raised five boys, often on her own, Loke said, "She's a survivor. She can eat anything. Which is good to know. And I think she's teaching all her kids that. That is really good. Cause some people, [they say] 'Oh no, how can you do that?' When you hungry, you can. When you gotta feed your family, you can. Just know how to prepare it, then everybody will learn to eat it."[35] Harvesting skills provide not just *'ohana*-level self-sufficiency, in which each family can fend for themselves, but collective security, even in times of disaster.

'Upena (Net): Surviving Natural Disasters

> *I can just visualize it yet, how the water, when it first came in, it rose all the way up. Good thing it didn't come where we were, but it came high enough that it went right up the river, over the bridge. I watched. I really was so amazed at how the wave carried that whole slab and pushed it back . . . about at least sixty feet from where the pilings are. And my brother-in-law's house . . . was lifted and went floating down the river and up the river, down the river until finally it disintegrated into little pieces.*
> —Nani Paik Kuehu, 2015

The ability of Halele'a fishing families to live off the land, share, and help one another not only built strong communities, but also helped these communities endure and retain their character even through natural disasters. Between the end of World War II and the turn of the twentieth century, seven hurricanes and two tsunamis made landfall on the north shore of Kaua'i. Families were able to recover from these disasters because they were accustomed to eating and living simply, not needing much. They could feed themselves from the ocean, rather than relying on stores and, through sharing, sustain their neighbors as well.

The worst of these natural disasters were two tsunamis that hit within a decade of each other, in 1946 and 1957. When the 1946 tsunami hit on April 1, early in the morning, there was no warning system in place, and many families were sleeping in their homes. Aunty Annabelle Pa Kam, age thirteen, was just waking up: "When I talk about it, I still can see it. . . . I was brushing my teeth. Our bathroom faced the river and our bedroom faced the river too. And my Tūtū was kneeling down, saying her prayer, and I said, 'Wow, this big wave coming inside,' just rising up. I said, 'Tūtū! Tūtū! Look at that big water coming in!'"[36] Annabelle's brother, Chauncey, was sitting at the back of the house by a door facing the hillside. When she yelled, he thought, "Yeah, April Fools. . . . Then, sure enough, water came rushing under the house, and a bunch of chicken eggs tumbled out." Their Tūtū took one look at the ocean and said,

> "Run, get out of the house and run." So we ran, behind the house [to] the mountain going up. Oh my gosh, that wave just came. And when it receded, [Tūtū] went, "Go—don't stop, don't look back, just go, go, go, go." . . . When I got to the top I looked back, the [bay was] dry all the way out . . . so dry.[37] And then, all of a sudden, this thing came back again. It's not a wave, it's just rushing, rushing. As she rush—she just builds up like that. So swift, so strong, and it just grab, . . . and you know, the telephone pole—only can see this much [holds out her hands to indicate a span of two feet]. That's how high it was.[38]

Kalihiwai during the 1957 tsunami as the ocean sucked out of the bay, then returned. (Photos courtesy of Ralph Akana 'ohana and Linda De Bisschop)

Merilee Chandler, who also lived with her grandmother, a few houses down from the Pas, wanted to warn their neighbors when she saw the ocean receding, but her Tūtū said, "There is no time." They ran out their door and uphill on the highway, the water rising behind them.[39] Just across Kalihiwai River, in Hanapai, the first wave broke on Uncle Shorty Kaona's family's house while they were still inside. The windows kept the water out. They ran as fast as they could toward the hills behind their home, turning to watch the second wave take their home away.[40] Down the road, in Kalihikai, Mrs. Julia Paik was the first to awake. There was water in their yard and the ocean looked funny. She sent her mother-in-law with the children into the hills, then went to wake other families and help them to evacuate, saving many.[41]

Kōkua: Helping Neighbors and Family

From the hillside behind Kalihiwai Bay, families watched helplessly as the water receded again and again, leaving not a drop of water for a mile and a half out to sea. "When the ocean finally stopped going out, it came back in very rapidly, rushing with full force up the valley."[42] People were swept upriver along with houses. One man rode the roof of his house as it washed up and down the river, then grabbed onto a large tree and climbed it to ride out the rest of the waves.

In between waves, though, people took action. Linda Akana Sproat's mother hustled six-year-old Linda and her cousins out of their house and up the hill just before the second wave. Aunty Linda recalled standing barefoot in her pajamas complaining about the *kukūs* (thorns) on her feet. "And pretty soon we saw our house floating by."[43] The gas stove was still on with a kettle of water to make tea, while a kerosene lantern sat lit on the icebox. "We could see the lantern going and shaking . . . all the way up the river."[44] Uncle Wendell Goo, age four at the time, remembers being trapped in his house as it filled up with water. His mother pulled him to safety. Young Chauncey Pa, watching from the hillside, saw his friend in trouble:

> I saw my friend, Richard Ninomoto, crawling on the debris of our house which had floated to a stop on the road. He was trying to save himself, but he didn't move fast enough! It was very difficult to crawl over all of the broken boards in the water. On the next wave he disappeared along with our house, which tore apart in a whirlpool in the middle of the river. Richard was gone. I saw his father, a county fireman [returning from work], rush toward the bridge and call for his family.[45]

Between waves, though, families huddled on the road overlooking the other side of the bay looked down and saw a little boy stumbling along the shore far out at the point. Some of the men ran down to grab him. It was Richard Ninomoto, every inch of his body black and blue. He was hospitalized at the Kīlauea plantation hospital and survived.

Less than ten years later, in 1957, the second tidal wave hit Kaua'i. In Kalihiwai, Aunty Linda's grandfather opened his *lānai* (porch) doors and saw the tsunami coming. It was the time of year when the family's twenty-seven-foot fishing boat, used for *hukilau* surrounds, was pulled up on the shore and flipped over for repairs, including caulking holes with cotton and *'ulu* (breadfruit) sap. She and her husband, Uncle David, recount how her father saved the family's fishing boat and multiple other community members at the same time:

Aunty Linda: So grandpa tells daddy, "Oh you better go save our big boat otherwise we cannot *hukilau*." So daddy runs to the bridge, dives off the bridge, swims across all the way and gets the boat, turns it over.

Uncle David: Big boat you know! Heavy boat!

Aunty Linda: Twenty-seven feet! And then he turns it over and uses a pole to pole past the bridge and by then the next wave comes.

Uncle David: What's remarkable is, the boat just made it under the bridge, then the next wave came. You know if he was before the bridge he would have probably slammed into the bridge. He just made it past the bridge and then he rode the wave. The wave carried him on the boat, he rode the wave up [the river] on the boat. . . . And he was rescuing people as he went by. All of them floating in the water, and he would throw his anchor rope to them and they would hold on and he pulled them to the boat.[46]

One of the people Aunty Linda's father rescued was his own mother, who was washed a half mile up the river, wrapped in barbed wire from their cattle fence, and deposited in a *hau* tree. In these stories, there is no Coast Guard, fire rescue crews, or professional outside help. Just as in feeding and fishing, the community relied primarily on their ability to take care of one another, for both rescues and recovery.

Shattering Loss and Community Restructuring

Seventeen people lost their lives in the 1946 tsunami, fourteen in all of Halele'a and five in Kalihiwai alone, including Richard's oldest sister, and Mrs. Mitsui, a storekeeper known for her generosity and kindness to all. Kalihiwai was especially hard hit because there is no reef fronting the bay, "nothing to slow down the monstrous tide." The bay funneled the incoming ocean, which hit the river, backing more water into the valley and creating whirlpools. Every house but one was destroyed by the waves, along with all three of the neighborhood stores. Kalihikai and Wanini, fronted by a large coral reef, were not hit as badly. There was no loss of life in these areas, and though buildings such as the church were moved off their foundations, most were not destroyed.

Down the coast in Hā'ena, a low-lying coastal area nestled against cliffs, whole families were lost to the waves. Aunty Honey-Girl Ho'omanawanui was holding her baby sister, only a few months old. She held on tightly during the first wave, but the second was too strong. Aunty Honey-Girl says she will never forget the feeling of the water sweeping the baby from her arms. When the ocean receded again, she heard her older sister calling to her from a tree, telling her to climb up quick. They rode out the rest of the waves in the tree, watching as their home was swept out to sea. Her family lost three children that day. For months after the tsunami, her mother sat on a pile of rubble in their yard, where she thought her children were, crying and looking off into space. Brother and sister Violet and Tommy Hashimoto recall being washed into a field behind their family home. They had to locate their siblings and their mother, who was wrapped in barbed-wire fencing, her clothes ripped off, but alive.[47]

No one was killed in the 1957 waves. This time a warning system of sirens made it possible for families from low-lying areas like Hāʻena to evacuate before the waves hit. One area *konohiki* fisherman, Uncle Henry Gomes, is said to have helped the families of Hanalei reach safety on the headlands above, from which they watched the waves hit. Dozens of homes were destroyed along the north shore, with hundreds left homeless—many, for the second time.

Kalihiwai and Kalihikai were both once thriving settlements—Kalihiwai with three stores and a gas station and Kalihikai with two churches and over twenty families. Since the highway was moved after the 1957 tidal wave, both places are now located along sleepy dead-end coastal roads with no stores or services. Aunty Annabelle Pa Kam explains how having two tsunamis hit with such devastation within less than one decade prompted some Kalihiwai community members to relocate: "When that [second] one came ... it took the bridge, the road, and everything. So everybody said, not going to happen again, because they still paying for their houses [from the first wave]. The plantation, the manager felt so bad, so he released the lots up there [on the ridge overlooking Kalihiwai Bay]. If you own a property down Kalihiwai, then you can buy one up there."[48]

After the 1957 tidal wave, the largest landowners and employers of the area, Kīlauea Sugar Plantation and Princeville Ranch, offered families living on ancestral taro patch and coastal lands in low-lying areas of Haleleʻa the opportunity to buy property at discounted cost on neighboring ridges with county government aid. While some families retained their former properties, and others gradually resettled there, these ridge lots made it possible for community members to relocate together. The new settlements on higher ground contained some of the same neighbors, contributing to the collective survival of the community. Other families shifted to new areas—for example, from Hāʻena to Kalihiwai or Hanalei so they could be closer to medical care and located away from a bridge, in case another disaster came.

Because of the strong social networks within and among *ahupua'a*, as well as the availability of land for resettlement, Halele'a communities continued to comprise many of the same families, using and sharing the same natural resources long after the waves receded, despite relocation.

Kūpa'a (Steadfast): Community Reliance and Resilience

> *I guess the way we grew up, because we never have money, money nothing to us. You know, everything was hand-me-down. And I was happy to have all the hand-me-downs. We didn't need anything new. We learned survival. That's how, when [Hurricane] Iniki came, we could live. We didn't need anything. We could live off the land. And that's what I teach my children and my grandchildren. How to live off the land.*
> —Annabelle Pa Kam, 2015

Community members described their local-level self-sufficiency as critical to their collective recovery from the multiple, recurring natural disasters, including seven hurricanes that hit Kaua'i between 1950 (Hurricane Dot) and 1990 (Hurricane Iniki), along with both tsunami.[49] When asked how her family recovered from the 1957 tsunami, Aunty Nani Paik Kuehu said simply, "What other choice did we have?" She went on to explain, "Well, . . . we lived so simple that loss of life would have been more than we could handle, but we had what we needed, we had our family, we had our love for all the family, and we managed to just do what we did every day the simple way."[50]

Kalihiwai and Hā'ena families recall their neighbors showing up after the tsunami with clothes, blankets, food, and other needed goods. Multiple families moved in together, with as many as four families and thirty-five individuals, twenty-five of them children, sharing chores of carrying water, collecting firewood, and cooking. Friends worked together to repair each other's houses. Gary Pacheco described shared work groups forming in Kīlauea town after Hurricane Iniki:

> See, a lot of the homes were tin roof. So the young men went around and collected all the tin roof [scraps blown off homes] and they went one house at a time and temporarily made it so that the people could live in it. It would leak here and there, but basically they had a roof over their head. . . . They would have a group of maybe twenty young men, . . . one group collecting the tin, the other ones trying to straighten it out, the other ones setting it up at different locations, and setting it. And the ladies provided lunch for the group. They would stop, maybe at noon, and . . . they brought round tables, utensils, and

everything and fed the workers. And then kids come in the afternoon to bring refreshments like ice water, soda.[51]

This cooperative effort to restore homes reflects community mobilization for other collective endeavors such as *hukilau* or *lū'au* (feasts). In these tasks, work is shared and the outcomes benefited multiple families. The ability to share in work and to rely on one another, as well as on the surrounding land and sea, provided collective resilience. Halele'a and Ko'olau communities have endured many changes and multiple natural disasters that have ripped and rewoven their fabric, yet have maintained their essential character. As climate change brings rising sea levels, projected to inundate Halele'a's coastal communities with more extreme surf events, tsunamis, and floods in the near future, it is important to consider how contemporary communities of the area will survive. Though natural disasters have relocated and reordered communities of Halele'a, sharing of foods from the land and sea persists, keeping longtime community members connected to their home areas and to one another.

Lawai'a: Sharing in Contemporary Times

> *They know, if you come back with plenty [fish], you better give every single house over here. So I know that within our 'ohana . . . it is automatic to share when you come back. I don't know if they were taught this way, but that was just the way it was.*
> —Kainani Kahaunaele, 2015

> *When summertime come, and the akule, that's the only chance we have, [to] feed everybody.*
> —Hanalei Hermosura, *lawai'a* and father in his forties, 2009

The practice of sharing fish continues in contemporary times, both from collective surrounds and individual harvests, as part of a broader system of community sharing and exchange. In 2008–2009, I worked with *lawai'a* from the community of Hā'ena who wanted to understand catch and distribution patterns from the area's reefs.[52] Ten fishermen tracked the fish they caught and who they gave it to for a year and a half, including two summer fishing seasons and one winter in between.[53] For each catch, we recorded all of the people and groups to whom fish were distributed, where they lived, and their relationship to the fishers. We found that 25 percent of distributions went to fishers' immediate family for home consumption, while 75 percent went beyond their home (63 percent to other families and individuals, and 12 percent for parties and other community gatherings).[54] Parties to celebrate the first birthdays of babies, honor high school graduates, or celebrate the lives of elders would not be complete without certain Haleleʻa fish prepared in particular ways: for example, *nenue poke* or fried *akule*. In fulfilling their *kuleana* to share their catch and feed others, *lawaiʻa* also bring people together around cultural practices associated with these species, including harvest, preparation, and consumption on important occasions.

Kūleʻa: Shared Competence

> *I'm not the fisherman, my brother learned how to fish. The only thing I'm good at on the reef is pick up limu. I cannot fish, I cannot even see the fish in front of me.*
> —Kealoha Saffery, 2015

Through sharing and facilitating collective harvests and consumption, *lawaiʻa* also perpetuate strong social ties within the community, with individual gifts strengthening connections between people. Uncle Kealoha Saffery describes friendships grown from regular sharing between those with different skills: "When you know you can help the next person—they catch fish, I catch mountain pig and make smoke meat. Share. You don't have to go fishing [and] go hunting, it's where you come close friends . . . by sharing."[55]

Fishermen regularly stop to visit with and bring fish to the homes of extended family members, friends of older generations of their *ʻohana*, or

Kealoha Saffery, beloved cowboy, storyteller, and uncle of Haleleʻa. (Photo by Kim Moa)

people who have helped their families. One fisherman says he always gives *māhele* from every catch to a woman who watched his children when they were babies so that he and his wife could work.[56] Their children are now in high school; however, he continues to honor this connection through his gifts of fish.

Fishermen express pride at being able to give generously and feed their extended families, fishing helpers, friends, and community members. *Lawaiʻa* feel sharing is a way to show respect and gratitude to their teachers, by using the *kūleʻa* (competencies and gifts) they have been taught to enhance lives and better the community. Fishermen regularly share with elders from the area, including former fishermen who can no longer harvest for themselves, acknowledging teaching and mentorship as well as the respected status of *kūpuna* in the community: "People who fished before, they couldn't come down [to the ocean] how they used to do. So [my grandfather] always taught me to go and bring [fish] to them. And they [were] so happy. That's what I remember—they happy. Some of the old folks, they cry, you know."[57] As one forty-year-old interviewee explains, he never went to college and he does not speak his Hawaiian language, but he finds it rewarding to use his gift for fishing and feeding people to perpetuate his cultural *kuleana*.

ʻĀina Momona: Collective Abundance[58]

> On this rock formation by the pier, the old Hawaiians used to have lines and people used to hang squid and dry their fish over there. And in those days nobody stole anything. They respected each other. In fact they did a lot of sharing those days. You never went to the store to buy, you got fed by the ocean. Everybody depended. And we were blessed 'cause we had farmers around us and we had cattle people around us that shared what they had with the farmers and the fishermen. So people survive. And everybody knew that and they respected each other.
> —Loke Pereira, describing her childhood in Anahola in the 1930s and 1940s, 2007

Historical accounts of *ahupuaʻa* emphasize trade within the community, often between upland farmers and coastal fishermen,[59] a practice that continues today. Gifting of fish is part of the broader sharing and exchange of abundance, including items such as smoked pig meat from hunters, beef from ranchers, vegetables from gardeners, mangoes and avocados from backyard trees, baked goods, and other specialties. *Māhele* of fish also facilitates exchange of skills such as child care, net repair, flower decorations, or preparation of culturally important dishes for family celebrations. *Māhele* thus continues the *ahupuaʻa* function of distributing abundance, in terms of both natural resources and human skills, through informal sharing networks.

This system of exchange is not a barter system, in which items are given with the expectation of immediate return of something of equal value. Instead, individuals share what they have in abundance, knowing that, if other community members do the same, everyone benefits without counting or keeping track of who gives what or when. As father and fisherman Hanalei Hermosura explains, "You might not get rewarded in money, but there are other ways we trade, we trade food. That's what they did before. Eventually, what goes around comes around."[60]

Families share skills and food with those in need, until a time when they again have something to share too. As Hanalei Hermosura explains of the catch he shares, "Sometimes it's people that get 'em just as hard as you. So by giving, they no need spend for eat that night, they can spend another night. You can give 'em tonight, then they can eat for two nights." In the economic recession of 2008, fishers regularly took fish to families where one or more parents had recently lost a job or been furloughed. Sharing without expectation of return creates a system in which families can share whenever they have extra, creating enough to cover those who are unable to give because of loss of a job or family member. Sharing extends benefits of self-reliance from the immediate family to the community level, acting as a form of collective insurance. Through sharing, even those families without individuals who fish can withstand economic disturbances and natural disasters, building community resilience.[61]

Harvesting to Survive, Sharing to Thrive

> *So you feed on each other's talents, you share that with one another. That is the part we miss, in today's generation. They don't think that way now. They on the opposite side, they wanna take.*
> —Loke Pereira, 2007

> *The most important is sharing. Once you stop sharing, you lose that, you never get back nothing.*
> —Kealoha Saffery, 2015

> *What if economic responsibility was defined by the amount we shared, rather than the surplus amount monopolized and hoarded?*[62]

What do communities look like when wealth is measured not by what we have, but by what we give away? In these stories, local families work to cultivate community ability to share, and the health of the surrounding environment sustains this generosity. The ability to give is a sign of wealth and thriving. Enacting *kuleana* through sharing creates abundance by bringing a variety of foods and skills together to benefit all. Here, community members' unique talents are integral to a system of "voluntary authority and subordination that shifts between people dependent upon context and each individual's expertise."[63] Living from the natural resources of the immediate area provided roles for everyone, shared work, and security rooted in relationships with other community members, and with the land. "A [contemporary] subsistence economy emphasizes sharing and redistribution of resources, which creates a social environment that cultivates community and kinship ties, emotional interdependency and support, prescribed roles for the youth, and care for the elderly. Emphasis is placed on social stability rather than on individual efforts aimed at income-generating activities."[64] Security and possibility reside not in stockpiles, savings in the bank, or outside entities, but in the abundance of the land and people's ability to work hard together to sustain and harvest that abundance.

Today, in Halele'a and Ko'olau, things are shifting, as people increasingly move in from the continental United States who can afford to buy beauty and bounty for themselves. Between April 2010 and July 1, 2016, Kaua'i experienced a net in-migration of nearly three thousand people, an average annual growth of half of 1 percent, bringing the island population to just over 72,000.[65] On average, one new person moves to Kaua'i every day.[66] Newcomers see the abundance of Kaua'i, where tropical fruits dangle from trees, as idyllic. *Kuleana*—the hard work, relationships, and balance of giving more than one takes—on which such abundance is built goes unseen. In reality, bountiful lifestyles depend on a community of families who share the bounty of their varied skills and care for one another. Much of this work is unseen or not recognized as work, such as hours spent watching the movement of schools of fish. Yet, this work is nonetheless critical to community well-being, survival, and abundance.

As newcomers purchase lands at higher prices, area families are forced to move away, leaving key *kuleana* unfilled, fundamentally changing social relationships, disrupting sharing networks, and fragmenting communities that have endured for centuries. Geographic and social dislocation caused by gentrification is more subtle and, in some ways, harder to resist than sudden natural disasters. Today however, just as in the past, communities are resilient and continue to thrive in new forms and spaces by fulfilling collective *kuleana* to live from the land, work together, and share. Even if they now live far apart, sharing keeps people connected to the land and sea of their home and to one another.

Moʻolelo: Huakaʻi Koʻolau (A Field Trip in Koʻolau)

Twenty-five third- and fourth-grade students stand on a bluff overlooking the Koʻolau coastline, as white waves flatten on the reef below them. They stand beside a windblown ironwood tree bearing a sign with tally marks for the number of people who have drowned at this beach. Clusters of ʻā, red-footed boobies, make their way past, fighting the light winds between the kids and the horizon.

The kids are in my son's class at Kawaikini, a Hawaiian immersion charter school where ʻōlelo Hawaiʻi (Hawaiian language) is used as their primary language of instruction. I have kept my daughters out of school to join the field trip, knowing it will be the most educational use of our day. We stand behind the double line of students, looking out to sea. It is January, and I remind them to watch for whales. One fourth-grade girl raises her voice to begin an oli, then the rest of the students join in, their words honoring the

Kawaikini third- and fourth-grade students enjoying their huakaʻi along the coast of Koʻolau in January 2017. (Photo by author)

highest mountain on Kauaʻi and asking to enter this place of learning. Next, the boys from Anahola, the nearby Hawaiian Homes settlement, begin the first few lines of the Hawaiian creation chant, Kumulipo. The opening words tell of how night was born from the depths of darkness, and from night the coral polyp, one of the eldest of ancestors. The students are chanting their genealogy.[1] As the students finish, a huge wave crashes onto the reef, pulling water in its wake as it recedes, exposing the expanse of coral below.

Among the group are ten students descended from three different *ʻohana* of this area. Their families are all named in the recent lawsuits brought by Facebook founder Mark Zuckerberg to quiet title *kuleana* parcels on the lands he recently purchased for over $100 million in the nearby *ahupuaʻa* of Waipakē and Pilaʻa. Those named in the lawsuits have twenty days to respond, or else lose any legal interest in their family lands.[2] Their *kumu* (teacher) Lei Wann, calls out the names of each area family in turn, asking "Where are you, descendants of Lovell, of Kaleiohi, of Kauhaʻa?" "*Ma hea ʻoukou e nā pulapula o Lovell, o Kaleiohi, o Kauhaʻa?*" The kids raise their hands shyly, and she laughs, "The *kūpuna* [ancestors] are so happy today, the sun is shining down on all of us because you are here."

Along the way here, more cars than usual were driving the quiet rambling road that once served as the main highway. The kids and I passed a white GM Yukon SUV pulled over on the shoulder across from the graveyard, the only remnant of the school, church, and settlement that once stood here. A cameraman and smartly dressed woman inside conversed with a contractor who does heavy equipment work on Zuckerberg's land. I figured they must be the CBS news crew who had left a message on my phone the day before. Other community members had also received calls from CBS, Al Jazeera, and BBC asking for comment on Zuckerberg's lawsuits. I had phoned Lei to ask permission to invite them along on the field trip. But when I called the CBS crew back earlier this morning, I reached a producer in New York. He told me it would be better to talk directly with the reporter, but she was in the field on Kauaʻi and hard to reach. Not wanting to keep the class waiting or arrive late for the chant, I shrugged and hung up.

These issues are not what bring the students here today. Today's field trip is part of the students' yearlong study of their grades' assigned Kauaʻi *moku*, Koʻolau. They are studying corals and *limu* and have been snorkeling with the state's Division of Aquatic Resources island-based biologist to compare healthy and degraded corals within Koʻolau. They are getting ready to raise native seaweed for later outplanting. They have read stories of Koʻolau printed in nineteenth-century Hawaiian-language newspapers and have listened to oral histories of family and community members. And they are creating a website to share their learning.

After the students greet this place with their *oli*, they walk briskly down a red dirt trail. The plan is to traverse three different *ahupuaʻa* today and glimpse a fourth, Pīlaʻa. Most of the students have never been here before. The day is quickly getting hot. One of the girls, Kaleo, offers to carry the mesh bag of extra masks and fins I brought for the kids. She is already lugging her towel, snorkel gear, change of clothes, and lunch but cheerfully swings the heavy bag over her shoulder and falls into step beside me. Other kids hold hands with my four-year-old daughter as she happily heads down the trail.

We pass *hauwī*, a medicinal plant sprouting from the red-dirt hardpan path. Kumu Lei stops to show them how to gather. She explains that some practitioners of Hawaiian medicine say to pick only with your right hand. Always gather when the sun is coming up, not after it reaches the top of the sky when the energy in the plants will also start to descend. This plant is good for aches and pains. A few of the kids ask to pick some for their mothers.

Small dirt paths intersect the main trail and lead through the bushes to the sand. Lei asks the students to wait quietly while she walks ahead on a path to check the beach. She tells the kids this place is known for *puhi* (eels), and she doesn't want to startle any. Lei returns, shaking her head vigorously and squeals, "*Aia ka puhi ma ka hale* [The *puhi* is home.]" Later she tells me that a man was on the sand at the end of the path doing his morning yoga in the nude. Lei moves the kids along farther down the trail till we reach the end of the beach.

Whenever introducing a new place for the first time, the teachers let the kids play and explore for a while. Some kids climb a tree to eat a snack. Others turn over rocks in the tide pools, building up the sides to trap baby *āholehole* fish inside. Some of the girls fan out to look for shells. Lei asks them to take home two pieces of trash for each shell they select. As they search, the girls name the varieties of dry seaweed they find washed up in the sand. Lei collects different corals, then calls the kids back together. She holds up each chunk and the kids identify them, calling out "rice, branch, lace!" English vocabulary interspersed with their Hawaiian chatter.

The kids pick up their bags, leave the sand, and follow the trail over a headland. Recently, longtime area fishermen have been stopped by security guards and threatened with arrest for hiking this section of the trail, which veers from the public beach and crosses onto private property for two hundred yards. When the trail drops back toward the sand, the kids drop their belongings and head along the coast. The beach is nearly empty.

The handful of people we do encounter are all here with a purpose. Two women stand in the treeline on a bluff above the coast, checking *mōlī* (albatross) nests for eggs. The kids stop and watch the pairs doing their mating dance and clacking their beaks. They crack up at the gangly bobbing steps that earned the nickname "goony birds." Some of the kids dance, imitating the *mōlī*'s movements. A red truck edges out of the trees pulling a trailer with a Jet Ski. It is the county's roving patrol lifeguards. Rather than sitting at one of the island's ten towers, this new team travels remote unguarded coasts to prevent drownings. At this beach, they warn snorkelers to stay away from the channels, with their sucking currents.

The kids reach the end of the sandy beach and step out on the rocks to look for seaweed. They learn to *ʻako*, gently pluck, tiny pieces of *limu*. They spread out to find ten different species to take back to their classroom and press for identification. One of the girls holds up a piece of invasive *Hypnea*. Kumu Lei describes how it can overgrow and carpet the reef, outcompeting the native seaweeds. The girl traces back to where she found it to check if there is more. It is a small population. Her friends fan out from the spot to make sure they pull all of it out. Kumu Lei encourages them to come back with their families to keep checking and weeding.

Lei points out a wet spot on the cliff. She thinks it is the spring one elder told her about, but there is not enough time today to check it out. As always, they will be rushing back to the bus, which has to be back by the final school bell to ferry students home. Lei wants the students to know of the many springs and streams of this area that have dried up since new landowners started digging private wells on the lands *mauka* of the coast.[3] I think of *kupuna* Uncle Henry Chang Wo, who taught Lei and me and so many others about *limu*, reminding us always that it needs fresh water to thrive.

As the kids start the long trek back down the beach, some of the girls spot three monk seals, one sleeping next to a *honu* with a four-foot-long shell. They learn that they can look up the number burned into the juvenile seals' pelts and track their whereabouts online. The girls don't want to leave the seals, but Lei reminds them that if they linger they won't have time to swim. The kids gather to bless their food, eat lunch, then jump into the ocean. Some of the boys cast from the rocks using fishing poles they've carried from home. Kumu Lei patiently untangles their hooks when they catch on the rocks. The intrepid kids put on snorkels and set off in search of fish, surfacing to holler the names of species they see. Noting the strength of the current, I position myself in waist-deep water and tell them they have to stay inshore of me. A gaggle of kids lolls in the surf, turning like monk seals. One asks if I think the current is what makes the coral so healthy, keeping the water moving and cool. I respond that it is likely that, and the lack of homes or buildings around here. We've seen only one all day, a new structure being built on the cliffs of the Zuckerberg property.

Before leaving, the students share an *oli* and hula to thank the place. (Photo by author)

Too quickly, it is time to leave, and the kids chant their *mahalo* before starting up the mile-long trail back to catch their bus. The kids are tired but hike fast, helping their slowest classmates by offering to carry their packs. We are hot again long before we reach the dusty parking lot where the bus is already waiting. The white Yukon is parked there too. The CBS news crew is preparing to launch a drone to fly over Zuckerberg's lands. None of the crew looks up as the kids rush to climb onto their bus. I linger for a bit and consider saying something to the reporter, but she looks busy. My own children are hungry and ready for a nap, so I load them into the car and head home.

The next night, CBS airs their report. There are breathtaking bird's-eye views of the coast we walked today, and a handful of protestors standing outside the stone wall lining the Zuckerberg parcel. I can't help feeling they missed the real story.

NOTES

1   Queen Liliʻuokalani translated the Kumulipo while imprisoned in her palace after she was overthrown.

2   Quiet title is nothing new in Hawaiʻi, but unlike this case, these actions are usually quiet. Sugar plantations have used this legal tool over the past century to acquire Hawaiian family lands. Notices run in the newspaper daily, and families who fail to see them or to respond in time lose their land without compensation.

3   Groundwater is regulated by the state's Commission on Water Resource Management (CWRM), which issues water use permits. However, CWRM only reviews wells in areas designated as "ground water management areas," defined by pumping levels that appear to fall below maximum sustainable yield (MSY). Recent research suggests that these MSY models are outdated and oversimplified, failing to capture potential impacts of climate change and the interactions between surface and groundwater. On an older island like Kauaʻi, with its broken-down rock and porous soils, water likely flows between streams and aquifers such that withdrawals from either affect both. Still, anyone who can afford the high cost of drilling can obtain a permit to build a private well.

Mo'olelo: Tūtū Makaleka Day

I grew up just up the hill from Kalihiwai Bay. We used to wait each afternoon for my father to come home from work and take us to the beach there, just as I take my children today. Not till after college did I meet a good friend, Kainaniokalihiwai Kahaunaele, who is named for that place, where her family are *kupa 'āina*. Since, I have enjoyed spending time with her fishing and musical *'ohana* in Kalihiwai, a whole new way to see and learn about this place I have loved throughout my life. In the fall of 2015 she invited me to attend their family reunion in honor of her great-great-grandmother, Tūtū Makaleka Kaluahine. This is a story of that day.

Kalihiwai Bay is calm this fall morning. White water crumbles gently onto the sand. Last week's north swell sent single waves walling up across the whole bay to crash with a boom on the sand named Kaihalulu.[1] The mountains stand watch, blue sky pulling clouds from their peaks as the day warms. Higher mountains behind stretch back to the interior of the island of Kaua'i and to Wai'ale'ale, long known as the wettest spot on Earth.

Today, sixty Hawaiian family members gather at the edge of the ocean. Some wear short *mu'umu'u*, colorful Hawaiian dresses in abstract plant prints. Others have sarong or pareu wrapped over bathing suits. Some of the men wear surf trunks and carry babies in their arms. Toddlers are digging in the sand. The older kids are already in the water, bobbing over the waves, diving under, washing onto the sand, then scrambling back out in squealing pairs.

Two women circulate among the group offering handfuls of white and pink plumeria blossoms, yellow and green variegated Pride of India leaves. A three-year-old child sticks the pointy leaves one at a time into a patch she's smoothed in the sand. These family members have flown from all over Hawai'i and even the continental United States to remember Makaleka (Margaret) Na'ea Kaluahine, great-, great-great-, or great-great-great-grandmother to most of the crowd. She was born here in Kalihiwai, in 1888, and raised her family in the small white one-bedroom home with a porch just across the road from the gathering. A taro farmer, fisherwoman, and quilter, Makaleka passed away in 1950.

Her home is the only one in Kalihiwai to have survived both tidal waves. The first tidal wave event, in 1946, picked the house up and spun it one way, then set it back down. The second, in 1957, turned it the other way. Family members joke that it now has the perfect view of the bay from the front

On Tūtū Makaleka Day in October 2015, Tūtū's descendants visit her Kalihiwai home, where many in this picture were raised. Two of her surviving grandchildren, both in their eighties, are in this photo. At the time this picture was taken, the family had not set foot on the property since losing it a year earlier. (Photo by author)

porch, and sits "just exactly where God wants it." Tūtū Makaleka's daughter, Margaret Kaluahine Pānui, affectionately known to her 'ohana as Wāwā, raised many of the family members here today in that house.

Earlier this morning the family met at Tūtū Makaleka's previously un-marked grave. They honored her memory with a new gravestone that reads, "Our beloved Tūtū, Me ke aloha pumehana." Our beloved grandmother, with warmest love. Now, they are at the bay across from her home, their family home, lost to a forced auction just a year ago, to remember those of her descendants whose ashes are scattered in the waters of the bay. One at a time, brothers and sisters, cousins and children, step forward to tell stories of their loved ones who rest here. "Margaret, you talk about your mom," calls out someone in the group. The rest of her sisters and brothers are up at the tent on the grass, preparing for lunch, so Margaret shyly steps toward the ocean and faces the group. The waves foam around her ankles.

"My mother named me Margaret for her mother, Wāwā, and grand-mother, Tūtū Makaleka. Mom was always taking care of everyone else. She wanted the family to be happy and together." "Like you," one of her cous-ins calls out. Margaret pauses. "One more thing. When I was a child I did not want my name. I told my mother, no one else is named Margaret, so old-fashioned. Today I finally understand why I have my name, and that it is a *kuleana*." She steps back into the crowd from hugs to those nearby.

Someone calls out, "And that red hibiscus in your hand, is so your mother. She always wore them in her hair."

When stories have been shared of each person in turn, the *ʻohana* steps into the sea together. Some people wade, others dive in. They lay their flowers and prayers on the surface of the shining sea. The flowers will float out to the bay and join the current toward Kalihikai or ride the waves to shore, spiraling in the shore break with the children all day long.

At the end of Tūtū Makaleka Kaluahine's family reunion, I cross the street with her descendants to visit her house. This house, their family home, has sat empty for the past year since the auction. Today, the family is returning for a short while. I located the woman now caretaking the home and asked if the family members could visit the house as part of their reunion. The adults all wear *ti* leaf lei made by a few of the older children the night before. At the house, Kainani plays "Aloha Kauaʻi," a beloved song of aloha for this island. Three family members dance the hula. The younger two women make their way gingerly up the collapsing steps to dance on the porch of the house facing their family. Aunty Merilee Chandler, in her eighties, stays on the grass facing them. She is Margaret Kaluahine's last living granddaughter. For the reunion, she brought the Hawaiian quilt her grandmother made at her birth. In the beautiful style of older dancers, Aunty Merilee hardly moves her feet. Her hands and her expression tell the story as she dances for the home her grandmother raised her in, for her grandmother, for Kalihiwai.

Aloha no o Kauaʻi
Luana, hoʻokipa malihini

Love for Kauaʻi,
this place of relaxation, welcoming to visitors

When the *mele* (song) is *pau* (finished), each family member, ages eight months to eighty, approaches the house. They drape the lei they are wearing and carrying on the porch railing, over the doorframe, around the empty, staring windows. My daughter and one of Kainani's nieces of the same age place a lei on the doorknob together. Family members filter back across the street, wiping tears from their eyes. Their lei bedeck the house, stirring in the breeze, right where they left them, over a year later.

NOTES

1   *Kaihalulu* means the rumbling sea. Even a small swell rushing against the cliffs outside this bay echoes at night through all the area homes, including where I grew up on the plateau above.

CHAPTER 5

Kīpuka

Kuleana to Land

*We always use the word kuleana to refer to land, but kuleana
is really your responsibility to that land.*
—Gary Smith, Kīlauea, 2016

*This is a sad situation. For me, I can see [my ʻohana] there for at least my
lifetime, you know. I'm OK. But eventually the parcel will be gone. That's the
heartbreak. And I think it's happening everywhere right now. All the Hawaiians
losing land, just out of—we just don't know how to deal with it. We didn't have
to deal with all this kind of stuff—we gotta deal with it now.*
—Haunani Pacheco, Wanini, 2015

*I tell my cousins, keep going. Bring your families. This is where your children
should learn to swim. This is where they should catch their first fish and play
volleyball. Because then they will have a sense of connection and responsibility, even if they live someplace else.*
—Kainani Kahaunaele, Kalihiwai, 2015

Kīpuka

Kīpuka are patches of forest that remain standing after a lava flow. The lava
separates into two streams, then the flows rejoin, leaving a stand of trees
surrounded by an expanse of fresh black rock. Seeds from the *kīpuka* spread
to the cooled and hardened lava. First the tendrils of ferns, then other native
seedlings, then trees gradually root into the lava and grow new forest surrounding the older, original stand. Rural areas of Hawaiʻi, such as the coast
of Haleleʻa and Koʻolau, are cultural *kīpuka*, "places where Native Hawaiian
culture [has] survived dynamic forces of political and economic change
throughout the twentieth century."[1] Often located in remote areas along
winding two-lane roads, cultural *kīpuka* have continued to be inhabited by
noho papa—Hawaiian families who have lived on the land for generations,
perpetuating cultural practices such as fishing and farming to feed their
families from the *ʻāina*.

Because of Hawai'i's history of dispossession, it is rare today to find Hawaiians living where their ancestors did, or to find those with genealogical ties and knowledge of an *ahupua'a* continuing to dwell there. Yet, throughout Hawai'i, and in Halele'a, *noho papa* families continue to be associated with the distinctive practices and character of communities. When asked, "Who is the community?" of any given *ahupua'a*, most longtime residents of Kaua'i's north shore name the same three to five *noho papa* families.[2] This chapter shares struggles of families of Halele'a and Ko'olau who have remained on ancestral *'āina* a century after the Māhele, despite ongoing commodification of land.[3] Multiple historical means of dispossession continue, including burdensome property taxes, shrinking access within *ahupua'a*, and legal manipulation to target Hawaiian family lands for development, along with newer threats such as termination of long-term leases, difficulties resolving interests of many descendant owners, privatization of vistas and other culturally important sites, increased recreational use, and competition with the wealthiest of global buyers.

Despite these struggles, Hawaiian fishing families, many of whom can no longer live in their ancestral homes, continue to find creative ways to resist dispossession and exercise *kuleana* that come with being of a place. I examine these efforts in light of the many Hawaiian words and proverbs referring to the people of a place. Together, ancestral language and contemporary actions illuminate *kuleana* that come with sustained connection to land: continuing genealogical ties, maintaining presence, perpetuating in-depth knowledge, caretaking, eating, and feeding. While community actions do not negate ongoing loss and injustice, these stories do offer possibilities: to restore lost connections, grow new ones, and build models that emphasize responsibility and caretaking of lands and resources, rather than ownership.

I ka 'Ōlelo no ke Ola (In Words There Is Life): Insights into Hawaiian Relationships with *'Āina*

Hawaiian language has a variety of words that identify the different relationships people have with a place, as well as their associated responsibilities. Some words describe families who have lived on particular lands for generations (*noho papa*) or whose ancestors are buried there (*'ōiwi*), while other words, such as *hoa 'āina* (friend of the land) and *kama'āina* (child of the land), can include those without genealogical ties to the place where they were born or reside. These varied relationships to land each carry distinct *kuleana*, expressed through sustained connection and care. Different words refer to those who reside on the land, get to know it well, cultivate and protect it (see appendix A). No Hawaiian words describe relationships to land that are established immediately on the basis of wealth or purchase. Descriptions of land emphasize that its spirit and life depend on *pono* (balanced and just) actions of the people, sustained over time. Hawai'i's state motto, uttered by King Kamehameha III in 1843, is "*Ua mau ke ea o ka 'āina i ka pono*" (The life of the land is perpetuated in righteousness).[4] For Hawaiians, the life of the people and the life of the land are intertwined.

History of Dispossession

> *Koe na'e ke kuleana o ke kānaka*
> Reserving the rights of the people[5]

Prior to Western contact, all *'āina* (land) in Hawai'i was held in trust for *akua* (the gods) by the *ali'i* (ruling chiefs and chiefesses), who were responsible for allocating land to all of the people.[6] The word *kuleana* described the "plots of land given by the governing *ali'i* of an area . . . to an *'ohana* or an individual as their responsibility without right of ownership."[7] No one could own land, source of all sustenance, manifestation of deities and ancestors. Though the *ali'i* of an area frequently changed, with land redistributed among the chiefly class, *maka'āinana* families largely remained constant, passing on the lands they cared for to their descendants.[8]

There were no land titles in Hawai'i until Kamehameha III privatized land through the Māhele[9] and the Kuleana Act, between 1846–1855. Some scholars argue that Kamehameha III designed the Māhele and the Kuleana Act to reflect Hawaiian land practices and secure tenure for native families within an imposed private property regime.[10] The king chose to apply the same word, *kuleana*, to the lands awarded to Hawaiian families—including home sites, *lo'i kalo* (taro patches), salt pans, and even *'āina kai* (ocean patches)—suggesting that the introduced Western concept of ownership did not change the fundamental Hawaiian view of the relationship between people and the *'āina* they tended.[11]

However, the concept and process of awarding ownership under the Kuleana Act was very foreign. *Maka'āinana* had to register for ownership of their *kuleana* lands, pay to have that land surveyed, find two other native residents as well as one "foreigner" or non-Hawaiian citizen to testify in support of their claims, and, if awarded, pay taxes.[12] Hawaiian *maka'āinana* were awarded less than 1 percent of all land in Hawai'i, with less than 29 percent of the eligible Hawaiian population actually receiving title to land.[13]

Though less than 1 percent of all *'āina* was awarded to Hawaiian *maka'āinana* in the small parcels known as *kuleana*, another 49 percent of the land was kept with the Hawaiian Kingdom government in trust for allocation to future generations of Hawaiians.[14] However, as control of governance shifted from the Hawaiian monarchy to American business interests, non-Hawaiians increasingly acquired government lands for private profit.[15] The Māhele changed the system of land ownership, but the change in governance and, ultimately, the overthrow of the monarchy by the United States, closed off important intended means for Hawaiian families to obtain formal title to land, leading to dispossession of the Hawaiian people.

The minority of Hawaiian families who were awarded *kuleana* lands received only quarter-acre house sites and agricultural plots that were in cultivation, not fallowed, at the time of the land survey.[16] These farm plots tended to be smaller than half an acre, though historically, families used and cared for much larger areas, *mauka* to *makai* (mountain to sea).[17] *Ahupua'a* residents once accessed forest resources such as timber for canoes and homes, medicinal plants, freshwater springs, farmlands, reef areas, and nearshore waters for fishing and gathering seaweed within established *palena* (boundaries).[18] While these *palena* sometimes corresponded with the boundaries of an *ahupua'a*, they were often smaller or larger, adapted to ecological conditions that shaped the productivity and availability of resources. Though irregular and difficult to map, in the Hawaiian mind, the *palena* within which a family could gather were clear, known, and understood by all who lived from the land. New systems of property ownership ignored historical boundaries of land use and the communities formed within them.

Though new ownership models were foreign and conflicted with native concepts of relationship to land, Hawaiians adapted by applying these models in ways that perpetuated *kuleana*.[19] Hawaiian families formed *hui kū'ai 'āina* (land-buying associations) to collectively buy back lands not awarded to *maka'āinana* through the Kuleana Act.[20] Local *hui* purchased four *ahupua'a* on Kaua'i, including neighboring Wainiha and Hā'ena, within Halele'a.[21]

In 1877, the Wainiha Hui raised $5,500 to purchase their entire *ahupua'a* from landowner and sugar company Castle & Cooke.[22] The Constitution of the Wainiha Hui awarded five acres of land to each member for homesites and agriculture, and allowed all members access to all lands of the *ahupua'a*.[23] Members could run no more than forty head of cattle, at a cost of ten dollars per animal after the first ten. Families who could not care for their lands could sell or surrender them to other Hui members, but not to foreigners.[24] The Hui met twice per year, and all members were required to attend, "except those who have good reason [a real problem, such as illness] presented to and approved by the association." Conflicts between Hui members were to be presented to the association before going to district courts. These guidelines preserve essential features of the Hawaiian land tenure system, including protected use of the entire *ahupuaʻa* by area residents, local-level decision-making, and *kuleana* to collectively care for and cultivate the land.[25]

The neighboring Hāʻena Hui endured through 1955, when two wealthy whites who had purchased shares from heirs of the original Hui members filed suit to partition.[26] This Hui maintained native communal land ownership a full century after the Māhele throughout the *ahupuaʻa* of Hāʻena. Today, descendants of Hāʻena Hui members, most of whom no longer live in Hāʻena, continue to care for former Hui lands.

Loss of Access

> *Hana ʻiʻo ka haole!*
> The white man does it in earnest![27]

> *Hawaiians were generally easygoing and didn't order people off their lands or regard them as trespassers. When the whites began to own lands, people began to be arrested for trespassing and the lands were fenced in to keep the Hawaiians out.*
> —Pukui 1983, no. 455

Privatization of land in Hawaiʻi created new boundaries around measured individual parcels, which accorded with Western ideals of ownership—discreet cultivated plots such as *loʻi* (taro patches)—while failing to recognize existing rights to the expansive diversity of ecosystems that Hawaiians actually used. Access refers to the "ability to benefit" from a natural resource, whether or not an individual has a formal property right to do so.[28] Means of access can include identity, such as genealogical connection or social relationships allowing one to gather in a given *ahupuaʻa*; knowledge of how to gather and use a resource; or possession of technology, such as a specific type of net needed to fish.[29] Changes in property ownership happen quickly, but changes in access can occur more gradually, as with access to lands within Koʻolau and Haleleʻa. Because new owners held large tracts of land for agricultural ventures that largely employed local residents, the people of Haleleʻa and Koʻolau managed to maintain access to most of their historic gathering areas for another century after the Māhele.

Early foreign settlers, including missionaries and traders, obtained leases and land deeds from the Hawaiian Kingdom to start agricultural operations on large parcels in Haleleʻa and Koʻolau. Some of these first agricultural ventures included a British sea captain's coffee plantation in Hanalei Valley in 1842; Scotsman Robert Wyllie's early attempts (1862–1863) to grow sugar on the Princeville Plateau; ranching and then sugarcane cultivation on the flatlands of Kīlauea by American whaler Charles Titcomb, starting in 1863; and a ranch on the Princeville Plateau founded in 1896 by A. S. Wilcox, a son of the first missionaries to Hanalei. After the overthrow, these entrepreneurs were able to purchase lands they had previously only been leasing, buy new lands, and expand their operations with secure land title. Two of these agricultural ventures, Titcomb's Kīlauea Sugar Plantation (1870–1971) and Wilcox's Princeville Ranch (1896–1969), employed workers from most of the families in Halele'a and Ko'olau for over seventy years. Working for these operations provided worker housing as well as continued access to area lands.[30] The plantation and ranch allowed community members to cross their lands to reach coastal fishing areas, hunt pig in the mountains, use plantation

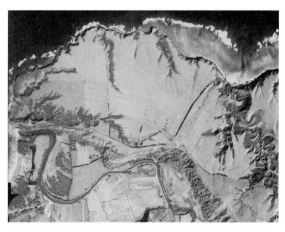

Kalihikai and Hanalei in 1950, showing Princeville Ranch lands. (Courtesy of Aurora Kagawa)

This image shows the same area of Kalihikai and Hanalei in 2013, including the growth of resort units and golf course in Princeville. (Courtesy of Dominique Cordy)

reservoirs to swim and fish for introduced bass and catfish, and pasture small herds of their own cattle on ranch lands to have ready access to beef.

When Kīlauea Sugar Plantation closed in 1971, its four-thousand-plus acres of land were partitioned by three realtors and sold, primarily to investors from the continental United States. Princeville Ranch sold its lands to a Colorado oil company, Eagle County Development Corporation, in 1969, which converted 995 of its eleven thousand acres to urban use for developing hotels and condominiums. Large tracts of both plantation and ranch lands were gradually sold, divided into smaller pieces, and sold again, creating a mix of small farms and luxury estates across Halele'a and Ko'olau.

New owners, most from the continental United States, have gradually fenced and gated their properties, precluding access by local residents. These days, pigs frequently run along the highway, perhaps because hunters have less access to mountain areas to control their populations. Families who can no longer use lands to ranch have become more reliant on fish as a source of protein.[31] Though public access along the shoreline is protected under Hawai'i's constitution and public trust doctrine, fishing is also becoming more difficult as new owners build on once-open coastal lands and close long-used trails that cross their lands to reach fishing grounds. A new echelon of wealthy owners moving into rural areas in the twenty-first century continues to close off multiple mechanisms of access to *'āina*, historically a source of both survival and abundance.

Anini: Kalihikai and Wanini

Research students and I conducted in 2015 offers an illustrative example of gradual, historical loss of ownership and use of lands on one two-mile stretch of coast in Halele'a, Kaua'i, known as Anini. Land ownership and demographics of the communities of this area changed substantially from the 1930s to the present. The area called Anini[32] actually encompasses three *ahupua'a*—Kalihiwai, Kalihikai, and the *'ili* (subdivision of an *ahupua'a*) called Wanini—within the *ahupua'a* of Hanalei. In the 1930s, twenty families lived along this coast, most Hawaiian, at the foot of coastal bluffs.[33] In the 1960s, after both tidal waves, a few families had moved away, though most remained and rebuilt their homes. A handful of new Caucasian immigrant families from the US mainland moved in, and twenty-five families were in residence into the 1980s.[34] All these families lived on the *mauka* (mountain) side of the road, across from the beach. People who grew up at Anini, between the 1930s and 1980s, describe the place as calm and tranquil. Everyone knew one another, and there was so little traffic that residents said one could sleep on the road undisturbed.

Aunty Annabelle Pa Kam, Haunani and Gary Pacheco, and Gary Smith talk story with graduate student Taimil Taylor during interviews at Kalihikai in March 2015. (Photo by Ron Vave)

In the 1980s, Princeville Development Corporation began to sell open lands along the ocean side of the quiet coastal road.[35] This land was intended for sale at low cost to the County of Kaua'i for a park as a condition for development of a second golf course on two hundred acres of former pastureland on plateau lands above Anini.[36] However, as one interviewee explained, "the county waffled on their purchase of the land. . . . Sylvester Stallone came in and bought it."[37] In 1988, Stallone, an American actor famous for his roles as Rambo and champion boxer Rocky, built the first house, with four bedrooms, 6.5 baths, and 4,807 square feet of living space, on the ocean side of the road, along with a nearby polo field. This first luxury home paved the way for neighboring coastal development. One local interviewee, whose family is from the Anini area and who was working for Princeville Development Corporation at the time, describes trying to buy one of the new coastal lots:

> All this place over here was *hau* bushes, *kamani* trees. That's all it was until Princeville sold it, this whole front part. I was working for Princeville, see? So when these properties came up, my girlfriend and I went to see if we could get one. By that time, [the company] CEO had already promised all these lots to all his friends [from the mainland]. We had no chance. Was $70K for one

place. We couldn't even touch it—it was already gone. [The lots went to] the guy who did [the CEO's] drywall, the guy who built his house.[38]

By 2012, the number of houses at Anini had grown to eighty-four, but only six qualified for homeowner exemption status as residences.[39] The other seventy-eight houses appear to be vacation rentals and second homes. Of the twenty Hawaiian families that lived along the coast in 1940, twelve continue to own their lands, but ten are currently at risk of losing them, and only two continue to live there.[40] Most lots at Anini are valued at over one million dollars.[41] One couple, both in their early seventies, who grew up in the area describes the change,

> Carol Goo: Just imagine all these families that used to live [here] and now it's all different . . . not even a handful [of local families].
> Wendell Goo: Now Anini is all millionaires.[42]

In fewer than seventy-five years, this quiet fishing community was transformed into a luxury beachfront resort destination, with most homes empty, few children growing up here, and only two Hawaiian families remaining.

Aunty Nani Paik Kuehu of Wanini helps Kauʻi Fu check her map made using Google Earth to record Aunty Nani and her sister Carol Paik Goo's recollections of the families who used to live along the coast of Kalihikai and Wanini. (Photo courtesy of Taimil Taylor)

Ongoing Forms of Dispossession

Local people are just getting priced out.
—Sterling Chisolm, 2015

The last time I came home, I went to Anini to say my good-byes, thinking that would be the last time I could ever go home. . . . I walked off the stairs and walked down the beach, I wanted to cry, thinking that the next time I come back I would just be a stranger looking into someone else's home.
—Maile Aniu Piena, age forty-five, raised in Wanini

Ongoing historic means of dispossession are becoming harder to resist because of growing demand from increasingly wealthy global investors. The coastal lands of the north shore of Kaua'i are some of the most coveted real estate in the world. Many of America's, and the world's, rich and famous own or have owned property within the communities described in prior chapters of this book, including Hollywood actors Ben Stiller, Julia Roberts, Sylvester Stallone, Bette Midler, basketball star Kareem Abdul Jabbar, the founders and early employees of Pixar Animation Studios, Facebook, GoPro, E-Trade, and eBay.

Though one of the most isolated land masses on Earth, Hawai'i is just a five-and-a-half-hour plane trip from California, making it a favorite place for millionaires emerging from the Silicon Valley dot-com boom to invest in real estate. These individuals build second, third, or fourth homes, to vacation in, rent as visitor accommodations, or sell at high return. With increasing high market demand, between 2008 and 2015 the average cost of a single-family home on the island of Kaua'i exceeded $700,000, with 43 percent purchased by mainland buyers. For the same time period, in Halele'a's largest coastal town, Hanalei, home prices averaged nearly $1 million, with 61 percent sold to mainlanders.[43] In the past few years, coastal properties in Halele'a have routinely sold for over $5 million; in 2017, twenty listings, on between .3 to 15.3 acres, advertised for between $1.98 to $70 million.[44] This new level of luxury exacerbates dispossession of local families through escalating property taxes, decreased availability of long-term rentals, targeting of *kuleana* lands through legal manipulation, increasing financial and emotional strain, privatization, encroachment, and loss of sources of sustenance.

Rising Values, Exorbitant Taxes

> *Striking ocean views are the hallmark of virtually every room in this custom home. Pristine beaches are just a seven-minute walk in either direction. Ocean conditions and premier surf spots can all be checked out from the breakfast table! Luxe extras include stained glass windows, Italian marble and a spa tub.... This beautiful home, in the midst of luxury estates, is in a friendly and low-key neighborhood. Truly, this beach community epitomizes the Hawaii experience.*
> —Sotheby's realty listing for a $1.8 million, three-bedroom home on 0.3 acres at Anini[45]

> *The degradation is being exacerbated by all the money that people have from elsewhere, who come here and buy and develop all this.... They're living the Hawaiian life, where the rest of Hawai'i is no longer living that.*
> —Gary Smith, 2015

Hawaiian and other local families are now losing their homes as they escalate in value. Property taxes on Kaua'i are tied to market value: $6.05 cents per $1,000 on a residential property.[46] Anytime land is sold, the price of the sale raises the market value of neighboring homes, and thus taxes. Taxes have dispossessed Hawaiian families of lands since the Kuleana Act first imposed monetary taxes on *kuleana* awards in 1850. Before that, taxes were paid in goods such as *'uala* (sweet potato), *kalo*, *kapa* (hand-beaten bark cloth), or *moena lauhala* (handwoven mats). Monetary taxes suddenly required families living trading and subsistence lifestyles to find ways to access money. Kumu Lei Wann, a Hā'ena woman in her late thirties, describes how her grandmother, Mrs. Chu, worked to pay taxes on their *kuleana* land throughout the 1980s and early 1990s by sewing Hawaiian quilts:[47]

> My grandmother, at ninety years old, she refused to give the *kuleana* taxes to my family. [After she passed] I inherited all of her books and her things; she wanted me to take care of them. I saw the prices of taxes she was paying over the years. She sold quilts for a living and she was using that money to pay for the taxes. When I really got to looking at the numbers, seeing what her social security was and how she was paying for things, it was hard. And you know *kūpuna* [elders], how they won't hand over things until the end. She wanted a better life for my dad and for us. And she sewed till she was blind. She couldn't see anymore. The last quilt she made, I didn't have the heart to tell her, "Tūtū, the lines are all off."[48]

Single individuals once managed, even at great personal cost, to pay taxes for an entire family. With today's rapidly rising property values, entire families working together can no longer afford tax bills:[49]

The tax [on our family land] is probably about twenty-two thousand [dollars] per year. My cousin in Hanalei, twenty-seven thousand. My girlfriend who had to move to Līhuʻe [the main urban area on Kauaʻi], hers was sixty-five thousand. How are you going to pay that? People who are retired don't even make twenty-seven thousand a year to pay tax. Something is definitely wrong with the whole system. It needs to be changed.[50]

Outstanding Hawaiian family tax bills in the Anini area range from $75,000 to $480,000.[51] One property's tax bill increased by $60,000 in 2015 because it was included in a new county residential investor tax class, based solely on its high value.[52] One forty-year-old mother of two expresses the pressure of knowing her family home will go to auction if they miss even one $5,500 payment on a back tax bill. "It is hard to want to save it, but not be able to, . . . financially."[53] Describing the struggle her family and others face, one seventy-year-old from Wanini said, "It's just too much. . . . They raise the taxes and pretty soon we lose our homes and everything we work for. We can't leave anything to our grandchildren because we have to sell. . . . We can't afford to pay the taxes. So what little we save, hopefully for our children, it'll be all gone. . . . We can't compete with billionaires. We're just simple people."[54]

Escalating Rents Cause Loss of Secure Homes

Escalating property values also inflate rents, making families who have securely leased or rented properties for generations also increasingly vulnerable to losing their homes. Many local families began to face evictions in the twenty-first century as landowners converted units on Kauaʻi to vacation rentals, in some cases to earn enough rental income to afford property taxes. With the help of websites like Vacation Rentals by Owner (VRBO) and Airbnb, vacation rentals have proliferated, not just in visitor destination areas but also in residential areas, town centers, and on agricultural lands.[55]

Some families face losing properties they have leased for over a century. Aunty Haunani Pacheco of Wanini shared how rents have increased on her family's 999-year homestead lease to a beachfront parcel issued to her great-grandfather in 1911.[56] In 1950, when Hawaiʻi officially became a US state, new laws permitted landholders to convert long-term leases to fee simple, giving them the option to sell.[57] This particular Wanini lease is one of many 999-year leases transferred from the territorial government to the state Department of Land and Natural Resources (DLNR), then converted to month-to-month. These leases provide no security and make it hard for families to make home improvements.[58] Without ever sending personnel to inspect her family's Wanini property, DLNR began increasing the rent based on surrounding property values, from $100 to $500 to $3,500 a month over the course of a year. This may have been in response to county tax assessments, increased from $7,000 to $18,000 per year between 2012 and 2013.[59]

To contest the new rates, Aunty Haunani's family took pictures of every house along the coast near their home, showing that most others were high-end vacation rentals. They argued that their humble one-story beach shack, with its tin roof and tiny porch, occupied year-round by family members, should not be assessed based on the value of surrounding properties. The family had a strong case, legal representation,[60] and paperwork to document the historic lease. They were able to negotiate a settlement that reduced the rent to $454 per month until the current family members who reside on the property pass away. After that, the family will lose the lease and DLNR can auction the property to the highest bidder. The family had to pay $90,000 in back lease rents owed because of the rate increases, none of which was returned to them. Aunty Haunani describes the stress of the legal battle: "They're expecting us to just walk away because the debt was $90,000. . . . They [DLNR] charged [my cousins] a lot more money than they were supposed to. But [DLNR] didn't go back and right the wrong, they did not give [my cousins'] money back. My cousin didn't have a car. She was walking to work, living like a crazy woman because she could not pay this debt. How can they do that?"[61]

Targeting *Kuleana* Lands through Legal Manipulation

Kuleana lands that were awarded to one individual in 1850 have hundreds of descendant owners today. Because any owner can sue to partition and sell the *kuleana*, debts or financial hardships faced by any single owner put all of the others in jeopardy.[62] *Kuleana* lands have certain rights associated with the original titles, including a house site, access through surrounding property, and sufficient water supply for taro cultivation.[63] These rights run with the land, whether or not owners are Hawaiian, making *kuleana* lands a target for acquisition by developers and large area landowners. It is possible to buy up shares from individual owners or pay outstanding tax bills, then quiet title the ownership through running an ad in the newspaper. To this day, quiet title ads for *kuleana* lands run monthly in Kauaʻi's newspaper. If families do not respond and find legal representation to contest the quiet title action within a set amount of time, they lose the land.

During the 1980s, Princeville Development Corporation acquired thirteen undeveloped *kuleana* parcels in the Anini area, within the *ahupuaʻa* of

Kalihikai. These parcels are small; most are less than a quarter acre. Three were located along the coast within what is now Anini Beach Park. The rest were former *loʻi* (taro patches) nestled along the meandering stream and *ʻauwai* (irrigation ditches) on the flats across the road from the park. The *loʻi* parcels were surrounded by Princeville Development–owned pasturelands and hard to reach by vehicle, making them difficult to sell. After acquiring the parcels, Princeville Development Corporation transferred these *kuleana* rights within their larger landholdings by subdividing their lands, redrawing the boundaries of the parcels so that they lined up along the roadway. Each *kuleana* retained its original size and the right to one house site, but was resituated to a location with easy vehicular access, with proximity to utilities and increased market value. Princeville Development Corporation then sold the lands, former taro patches converted to investment homes rented to tourists.

The cultural value and history of *kuleana* lands is documented more thoroughly than most lands in Hawaiʻi through century-and-a-half-old land deeds, surveys, and testimony by claimants and neighbors describing the parcels and their surroundings. Many of them have appurtenant water rights and some have burial sites.[64] At the time that the Kalihikai parcels were moved, in the 1980s, the Kauaʻi County Planning Department was very concerned that the movement of *kuleana* from one spot to another through the subdivision process was not simply an adjustment of boundaries. Through moving *kuleana*, "you lose the historical value of how things were at one point in time."[65] Today, few community members know these *kuleana* lands or the Hawaiian families connected to them ever existed. In their historic locations, many *kuleana* are landlocked in valleys, islanded within larger properties, and difficult to access, making them valuable mainly to Hawaiian families who descend from these lands or want to restore them to taro cultivation. The ability to relocate isolated *kuleana* parcels to more lucrative locations makes these lands attractive to developers and large landowners, raising their value so that these longest-held and last affordable Hawaiian lands are also increasingly out of reach for Hawaiian families today.[66]

Kuleana lands have become a growing target in more recent years, associated with development driven by the extremely wealthy. In another example from the Anini area, in 2014, the company that bought Princeville Development Corporation, Resort Group, unveiled plans to build 350 luxury homes and condominium units on the ridge above Anini, as well as create seventy-five "ranchette" lots on pasturelands across from the coast (including twenty in the same area the *kuleana* were moved from in the 1980s).[67] The project was to be developed in partnership with Discovery

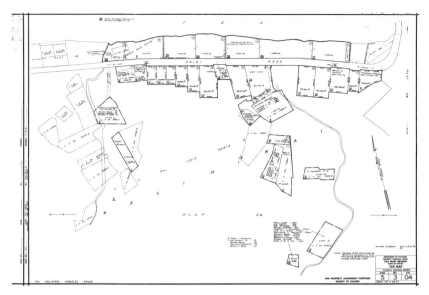

This map shows the original locations of house sites and *loʻi* (taro patch) *kuleana*, the faint irregularly shaped parcels along the stream on the left. The relocated parcels are the rectangles lined up along the mauka side of the roadway (or below the road on the map). (County of Kauaʻi Tax Map)

Land Company as part of its exclusive resort club model.[68] Prices were ten million dollars per lot, with multigenerational family membership fees of one million dollars per year. A preliminary marketing brochure proclaimed the project "a playground for princes then and now. And an opportunity for relaxed, country living, island style, that just a chosen few may call home." It invited buyers to "create a family retreat for generations to come" and "enjoy the array of amenities benefitting the new *konohiki*, stewards of the land."[69]

Hawaiian family lands were targeted as sites of such amenities. One long-time Hawaiian family's home across from the local park and boat ramp was slated to be a boating center and storage site for club members' watercraft. The Resort Group also began targeting *kuleana* lands at the end of Anini road below the development. The Resort Group had already acquired one of these *kuleana* in 2006 for $2 million, and in 2014 sold it to one of the development's investors, Reignwood International, the Chinese distributors of Red Bull, as part of a larger land sale for $16.9 million. In development plans, the land was intended to become a beach grill, club, and stand-up paddling and kayaking launch point for club owners and guests.

In May of 2015, Resort Group acquired a single share of a neighboring *kuleana* lot and paid off the outstanding taxes.[70] The county had moved to foreclose, and the property was set to be auctioned. The auction was cancelled once the taxes were paid. In this way, The Resort Group was poised to quiet title and acquire the land for $75,000 instead of its $1.7 million assessed value. To keep the land, the descendant owners, scattered across two other Hawai'i islands, California, and North Carolina, would have had to buy back the developer's share, pay the back tax bill, contest the quiet title actions in court, and find a way to keep up with future taxes. As Haunani Pacheco states, "That's what's happening to a lot of Hawaiians. Some of them just don't want to deal with the fight because they don't have the resources to do so or they don't know where to go, to do so."[71] Legal proceedings are lengthy, expensive, and difficult to win against corporations and individuals with access to tremendous capital, prohibiting many families from challenging such actions.

A highly publicized example of targeting *kuleana* lands for acquisition involved Facebook founder and billionaire Mark Zuckerberg. In 2014 Zuckerberg and his wife bought one-and-a-half *ahupua'a* in the district of Ko'olau for more than two hundred million dollars.[72] He then set about systematically attempting to buy thirty-four small *kuleana* lands within his seven-hundred-acre holdings to ensure a completely private retreat. *Kuleana* parcels have access rights, meaning that hundreds of descendant owners of these properties located within the *ahupua'a* could legally traverse Zuckerberg's property. Two parcels of less than half an acre sold for half a million dollars each in 2014, while a 1.07-acre parcel sold for just over six million dollars. These sales raised the value of all other *kuleana* in the region, along with their tax bills, making it simultaneously more profitable for remaining families to sell and more difficult to hold out as taxes escalated.[73] In 2016, Zuckerberg filed eight quiet title actions in Hawai'i State court containing adverse possession claims on around a dozen remaining *kuleana* he had not yet been able to acquire. With *kuleana* lands, it is a challenge to identify, much less buy out, the hundreds of descendant owners who hold shares, some of whom may not even be aware they are owners.

The quiet title and partition suits, filed on December 30, 2016, would force families to sell their shares to Zuckerberg or see their land auctioned at a public court to the highest bidder.[74] Hundreds of people named in the suit, many no longer living, and hundreds more of their descendants with interest in the *kuleana* parcels were given twenty days to respond in court and defend their claims to the land.

Quiet title actions usually proceed quietly, with descendants missing the advertisements posted in the paper, not realizing their land is being acquired. However, the Zuckerberg action was far from quiet. His lawsuits prompted multiple protests outside the newly constructed rock wall closing in his estate, and global media coverage by CNN, Al Jazeera, and the BBC. There were nearly ten thousand critical comments posted to his Facebook page.[75] Native Hawaiians and other concerned community members used Facebook as a tool to alert those named in the quiet title actions and their descendants. Twenty-five individuals met the court deadline to file. Those who spoke out publicly had no interest in selling, despite the high prices being offered for small fractional shares of the properties, such as a 1/256 portion.[76] One woman whose partial share amounted to as little as three feet by three feet stated that she would not sell for any price. Zuckerberg was surprised by the backlash, as he had been advised that he and his wife, Priscilla, were doing the right thing by "initiate[ing] the quiet title process to identify any other partial owners so we could also pay them their fair share," no matter how small their individual shares might be.[77]

In response to the outcry and quiet conversations with multiple community leaders, Zuckerberg and his wife decided to withdraw their quiet title lawsuits. In a letter to Kauaʻi's only daily newspaper, the *Garden Island*, published on January 28, 2017, Zuckerberg explained their decision:

> We understand that for native Hawaiians, *kuleana* [lands] are sacred and the quiet title process can be difficult. We want to make this right, talk with the community, and find a better approach.... Upon reflection, I regret that I did not take the time to fully understand the quiet-title process and its history before we moved ahead. Now that I understand the issues better, it's clear we made a mistake.... We love Kauaʻi and we want to be good members of the community for the long term. Thank you for welcoming our family into your community.[78]

Quiet title actions have historically been used as a means for large land-owners such as sugar companies to acquire *kuleana* within their larger holdings. However, quiet title is also a vehicle for Hawaiian families to prevent forced sales of their lands. Quiet title actions are a means to identify all potential owners, buy out those who want to sell, and consolidate shares with those who want to keep the property, making it less vulnerable to eventual forced partition and loss. In Pīlaʻa, one Hawaiian man chose to join as a party with Zuckerberg in the suits. Zuckerberg had already acquired shares of some of this man's four family *kuleana* and was in position to force partition and sale. Zuckerberg would finance the expensive quiet title action to enable the man, who had lived on and farmed the lands for over forty years, to pass on clear title and secure ownership of two of the parcels to his children, in exchange for giving up the two other parcels. Other

descendants of the man's same family were some of the most vocal in pro-
testing Zuckerberg's suits. This example illustrates the tough choices and
tensions Hawaiian 'ohana must navigate in finding ways to keep *kuleana*
lands within an imposed Western property system. Systematic targeting of
kuleana intensifies financial and emotional strain within families who continue to hold these lands.

Emotional and Financial Strain on Families

> *I know of a family, they had beachfront too, they gave up, they just walked away*
> *from it. They couldn't do it. Or they have so many people on it [as owners] that they*
> *can't come to an agreement about it, and nobody pays the tax . . . and they end up*
> *losing it again. But I think it [the system] was designed that way, to fail.*
> —Haunani Pacheco, 2015

Under Western private property law, *kuleana* lands are not collectively held
by families, but owned in partial shares—which may be as small as 1/512—
by hundreds of distantly related individuals. Keeping lands requires organizing descendant owners, many of whom have never met one another or
seen the land, and who live scattered throughout Hawai'i and the United
States. With the addresses of home and land owners publicly available on
the County of Kaua'i's real property website, realtors, most of whom have
themselves moved to Hawai'i from elsewhere, routinely mail glossy bro-
chures highlighting million-dollar sales of nearby properties and offering
similarly lucrative results. Owners find it increasingly hard to resist selling,
as both property values and tax burdens increase.[79] Any single owner can
sue to partition and force sale of the *kuleana* at public auction. Kainani
Kahaunaele describes how relatives raised on the continent, with little
connection to the land or 'ohana, forced auction of her great-great-grandmother's home at Kalihiwai (described in the opening of this chapter).
Some of these individuals had never actually visited these *kulāiwi* (lands of
their ancestry):

> Just last year, we had to let it go. It was a very big deal for our family. It took a
> lot of meetings and a lot of convincing. Because there were many owners, . . .
> [more] with every passing generation. It was our line, and we had our home
> there, but our land still belonged to others as well. Certain lines of the family
> didn't feel it worth their time. "Well, we aren't benefiting like you folks are"
> [they said]. So we had to let it go. Everybody's processing this differently. Some
> people are just so sad, some people are very angry. . . . I say, well, it is not worth
> the family fighting. . . . So, we are transitioning.[80]

Another *kuleana* owner expresses why she and her elderly husband, who have fond memories of family land at Wanini, might opt to sell after contributing to the tax bill for years, even though they don't want to: "He loves this place and it bring back lots of childhood memories for him. We are in favor of keeping the property, but if there is no participation, by everyone concerned, then we should sell it. . . . It's not fair to one or two people to take charge and bear the burden because it is hard. We know that from experience."[81]

Some families choose to give up lands in an effort to avoid arguments and conflict within the family, or out of respect for elders. As Kahaunaele explains, younger generations of her family respected the elder members' decision to let their land go. "We still have my grandmother's generation. We still have members of that generation and their designated representation, it didn't reach mine yet. It is still with the elders, so I just go, 'Okay, it is what you want.'"[82] Other families wait to resolve property issues to refrain from burdening elder generations. As one seventy-year-old shared, "It's too hard for my mother to have to deal with this again. You know, in her condition, it would be very hard. I need time, have to wait to settle this."[83] Multiple families have also lost lands with the passing of elder family members, sometimes because reverse mortgages had been taken out to cover debts or property taxes. In other cases, families cannot come to consensus on how to

Kainani Kahaunaele with her husband, Halealoha Ayau, outside her great-great-grandmother Tūtū Makaleka Kaluahine's home in Kalihiwai, November 2017. (Photo by author)

handle the property after a key elder passes, potentially losing their matriarch, family home, and gathering place all at once.

As much as they speak of loss of particular parcels of land, people speak about the loss of social and familial connections associated with that place. Llwellyn Woodward, whose ʻohana has lost both his mother's and his father's family homes at Anini, describes spending times when all his cousins would come home in the summer: "So when I was brought up, our cousins, especially summer time, they used to fly down from Oʻahu and every once in a while my cousins from the mainland would come and we would stay in the water *all day*. It was very special, it was safe. We never worried about TV and all that. I'm sad . . . that's [not] how my kids is brought up now, since we got out of Anini."[84]

Some families that have lost lands say it is harder to find places to get together and easy to do so less frequently. Kainani Kahaunaele described how their recently auctioned family home is integral to who they are as an ʻohana, saying, "This is where we know how to be a family." Her aunty elaborated, "All the things we do, the jobs we have, the *kuleana* we all carry, are because of the things we learned here."[85] As they lose family lands and are forced to move, families face changes in their cohesion and character. The character of their home areas changes with them, because the spirit of the ʻāina and its people are one.[86]

> Lele ka hoaka.
> The spirit has flown away.
> The glory of the land has departed.
> Also, the person is dead.[87]

Encroachment

> When you look up left on that hill, never in a million years did we think we'd have homes up there. . . . We never thought about the view up there. I have friends that service those houses, you know that, do the landscaping and the termites, and they take pictures of the place and I think, "Oh my gosh, how gorgeous," and at the same time it tears me apart. I mean, nobody is living up there. That is the tops of our hills, tops of our mountains, the big trees, our landmarks and our graveyards right there. . . . We never thought that could happen.
> —Kainani Kahaunaele, 2015

Encroachment is a slow shifting of the landscape, incremental, which makes it more difficult to resist. Visible signs on the landscape tell of shifts in human relationships with a place. Lava rock gates and vacation rental signs replace open yards where humble one-story homes once stood, signaling the transition from coasts as community sources of sustenance to private places of vacation and recreation. Quiet communities of families with comparable incomes have become bustling vacation retreats where luxury mansions dominate vistas as new owners mark their presence on the land. The character of coastal areas and people's ability to experience them has been transformed, making it harder for families to want to keep coastal lands, even if they can afford to do so.

Hawaiian families built their homes along the coast, set back from the beach, where there were often burials, or in the valleys, near water sources. Ridges were reserved for temples, and after the arrival of Christian and Asian settlers, cemeteries. Gifted fishermen were often buried at high points, on ridges and hills where they went to *kilo*, observing the movements of schools of fish in the ocean below. One of these grave sites is now located just downslope from a golf course hole. Area cemeteries are surrounded by luxury homes with swimming pools visible through hedges planted right up to the graves. Encroachment changes the experience of descendants visiting, transforming not just specific grave sites, but the tranquility and vistas that made people want to bury their loved ones in high places of esteem. Some families have erected headstones on formerly unmarked graves to ensure they can be found and cared for in the changing landscape.

Many interviewees described staying away from areas where they grew up, avoiding entire beaches and towns on the north shore, particularly Hanalei, because it is too hard to see how dramatically they have changed in a short

Aunty Inoa Goo Aniu with her daughter, Makana Bacon, and granddaughters, Analani and Keʻahilani Bacon, after Sunday service. (Photo by Anuhea Taniguchi)

time. One fisherwoman in her early sixties, Aunty Inoa Goo Aniu, told of
trying to return to the reef in front of the home where she was raised by her
grandmother and where she raised her own five girls, to go for *he'e*":

> We often went at night, and I knew it would be quiet then, so I tried to go back
> a few years ago. I hadn't been to look for *he'e* there for at least ten years since I
> moved here to Anahola [Hawaiian Home Lands]. But there were so many new
> houses with lights shining on the water, I couldn't see the reef to find the holes.
> Or the stars. There were always so many stars. I haven't gone back since.[88]

Through encroachment, families are losing not just fishing spots or grave
sites, but also the experience of going to these places to visit and harvest, as
well as the spiritual sustenance they provide.

Losing Sources of Sustenance to Recreation

> *Mom and dad used the land for food and that's what we wanted to*
> *teach our children. That's where we got our food, fished, got limu.*
> *Helped other families who needed help.*
> —Anonymous Wanini Community Member, 2016

> *In Hā'ena [I see] the desire to see some semblance of community they grew up*
> *with that was quiet and now so overrun by people playing in their resource and*
> *community members just want that old community back.*
> —Debbie Gowensmith, 2009

Visitors, their intentions and interactions with the land and sea, have also
changed the experience of those who access places for sustenance. Kaua'i
has a long history of tourism, but tourism has changed in significant ways
over time. Tourists today are self-guided versus hosted, with more tourists
venturing to more areas. Many buy land and become annual repeat visitors or full- or part-time residents. *Kama'āina* interviewees who grew up
in coastal communities across Halele'a from the 1940s to the 1970s describe
roaming empty beaches as children, beaches now crowded with visitors. A
community member in her late sixties states, "In my lifetime, look what I've
seen, from nobody to everybody, right?"[89]

In the 1950s and 1960s, after the tidal waves, visitors to Kaua'i mostly
toured the island with local guides. Through the 1970s, 1980s, and even into
the 1990s, visitors tended to stick to certain beaches, such as world famous
Hanalei Bay, consistently rated one of the top ten beaches in America. Today,
Kaua'i's 1.17 million annual visitors[90] stay in vacation rentals, time-shares,
bed and breakfasts, and campsites scattered along the coast, and venture on

their own using guidebooks, social media, Internet sites, and GPS to direct their rental cars.

Many of these "self-guided" resources direct tourists to dangerous and remote beaches.[91] Residents frequently have to warn visitors away from high waves and rocks, or even rescue them.[92] More than eighty people have drowned on Kaua'i in the past decade, over three-quarters of them visitors.[93] One woman shared her concern for the increasing number of drownings in her home area, Anini.[94] She said in talking to her father and older relatives, including one ninety-three-year-old uncle, no one remembered drownings at Anini until recently.

> My question is why [now]? Because the water conditions have not changed all that much. The only conclusion that I can come to is that there's no longer anyone living down here that knows the water, who might say, "You know, you might not want to go out there today, it's really kind of rough. The waves are ... going to pull you this particular way." So, for me it's that personal connection to place. When we're talking about preserving resources, the people are part of the resource.[95]

As Kaua'i's visitor arrivals increase, tourists, especially repeat visitors, increasingly seek remote, less-crowded destinations, until no place on the island seems untouched by a constant stream of people. In contrast, local families were raised to stay away from certain places within their *ahupua'a* considered sacred or unsafe. Off-limits places included spawning areas for fish, temple sites, caves or swimming holes known to house *mo'o* (water spirit lizards), and certain beaches prone to dangerous surf. "We never really ventured up the river. . . . Now they have kayaks and paddle boards, it is so much more accessible. But we didn't go up there, because we didn't have anything to do up there. . . . I mean it is a beautiful site, a gorgeous place to go take a look. But growing up, that is not something that was encouraged."[96]

Ko'olau and Halele'a families were taught to go places for a purpose, to gather food, learn, or take care, rather than simply to go. When visiting other places beyond their home areas, people spent time with local hosts, respecting their guidance on where to go and how to behave. Today, unguided tourists engage in dangerous and inappropriate behaviors, from camping in temple sites to performing yoga in caves to swimming with dolphins to jumping from waterfalls—all examples featured in media coverage over the past year. One younger community member, who is the *kahu* (caretaker) of the *heiau* (temple) for hula training located in Hā'ena, describes recent challenges of conducting ceremonies there: "Now you have to go check make sure nobody is up there before you start your ceremony. Before, you wouldn't really have to do that. The hippies would stay away, the tourists would stay away from those kinds of places. Now everybody's everywhere. So, trying to protect these places, because you know, if *I'm* getting irritated about it, imagine the place!"[97]

Fishermen frequently describe challenges of fishing in waters crowded with surfers, windsurfers, kite surfers, snorkelers, Jet Skis, stand-up paddlers, and swimmers. One longtime Haleleʻa fisherman, Wendell Goo, and his wife Carol, explain,

> Wendell: Before, very few people used this road. Mostly families who
>   lived in the area. If you meet one car, you lucky. But now, . . . 8:30
>   in the morning you go down there, [people] coming already.
>   . . . Before take your throw net, go to the beach, guarantee you
>   get something to eat. Today, might come home with white wash
>   [*laughing*].[98]
> Carol: You do come home with white wash.
> Wendell: [Sometimes] the whole place . . . all full, just like Waikiki.
>   More people, more disturbance.
> Carol: All kinds of ocean activities. Lots and lots of tourists.[99]

Wendell Goo of Kalihiwai and Carol Paik Goo of Wanini at student presentation of research, May 2015. (Photos by Taimil Taylor and Ron Vave)

In Haleleʻa, fishers give one another space, refrain from venturing out on the same patch of reef, and wait to make conversation until fishing is done to avoid disruption.[100] Today, fishermen about to throw their net on a pile of fish are frequently approached by curious tourists who scatter the school.[101] *Lawaiʻa* worry that the high volume of constant recreational use disturbs feeding and spawning patterns of fish.[102]

Loss of land has also led to loss of access to fishing areas. Fisher men and women now have to drive to their fishing spots, as they live farther away, rather than waking up across the street. Once-open coastal lands are now developed and gated, leaving only sporadic fenced-in chutes as public rights-of-way to the beach.[103] Aunty Violet Hashimoto, an expert fisherwoman in her early eighties, was turned away from the Hāʻena trail she had used her whole life to walk to the beach to look for *heʻe*. The new landowner told her she was trespassing on private property. Another eighty-five-year-old fisherwoman described a similar problem along the coast of Anini: "All the [new owners] . . . , they block off their place. Before, we used to go through yards, but everybody knew everybody. [If] we catch fish, then we give them."[104] Though rights of Hawaiian families to access historic fishing trails are protected by law, at least three trails along this coast have been closed or moved by landowners since 1990. Head fishermen once climbed tall trees and ridges to *kilo* (observe) schools of fish and direct harvests. These viewing points are now situated on private lands, off-limits, and important *kilo* trees have been chopped down. Loss of *ʻāina* compromises sources of nourishment, both the ability to provide food, such as fish, and to provide learning and replenishment.

Means of Resistance and Resurgence:
Maintaining Ancestral *Kuleana*

> But when our family is out of there and they say, "We don't want to go . . . it is too much traffic," that is when I say, "No, that is why you've got to go.". . . I'm trying to shift the mentality, because the more you stay away, the more you leave the spot open for someone else. I've seen in this community, for our family to maintain, they have to maintain a presence. They don't have to make a big ruckus. They just have to have a presence and, you know, clean up the driftwood. Pick up over here and over there. Say "Hi" to the neighbors. But the presence part is what we have to work on.
> —Kainani Kahaunaele, 2015

Despite ongoing dispossession through the changed character of areas and loss of ownership of family lands, Haleleʻa *ʻohana* continue to quietly resist through everyday practices that perpetuate relationships with the land and sea. Families find ways to maintain their presence and enact *kuleana*, including fighting to keep family lands, perpetuating genealogical connections by gathering as a family and caring for grave sites, continuing to fish and eat from their home areas, teaching future generations knowledge of these areas, and continuing to care for them.

These everyday acts, which some scholars call resurgence,[105] mirror ancestral *kuleana* expressed in Hawaiian language through the multiple Hawaiian terms referring to the people of a place. In the following section, words and *ʻōlelo noʻeau* (proverbs) that elucidate *kuleana* inherent in various relationships to land are interspersed with stories of families' actions to maintain connection (see table in appendix A).

Noho Papa: Continuing to Dwell on the Land

> *No ka noho ʻāina ka ʻāina*
> The land belongs to the one dwelling on it[106]

> *But how can I? My grandfather had sixteen children on this land.*
> *They were brought up in this place, my grandmother was very connected*
> *to the land. How do you give it up?*
> —Haunani Pacheco, 2015

Despite seemingly insurmountable struggles, many individuals and families refuse to surrender ownership of their land. They work ceaselessly to find ways for members of their *ʻohana* to *noho papa*, continue to live on and return to the land, and to protect this right for future generations. One family member or branch often takes on the burden of paying taxes for all the other descendants, though the payer may own only a 1/125 share.[107] Many contribute and sacrifice though they will never live on a property themselves. As one interviewee said, describing years of legal battles and work to keep a property, "It causes friction between families 'cause they're [other family members] thinking I want it, right? I have no desire. My cousins, they belong there, they need to stay there."[108] Some *ʻohana* turn the parcel over to one nuclear family branch or member, gifting or selling it at very low cost. As another interviewee explained, "I turned it over to my niece, 'cause they go down there fishing. I thought 'Oh well, since I am not using it, I might as well turn it over to them.' So [they] have it."[109] This approach simulates private ownership, allowing one party to make decisions, pay taxes, own the property outright, and access home equity, for example to send children to college. Yet, the property stays in the *ʻohana*.

Another family decided that no one would live on their parcel so that all could feel free to gather there. One of the family members explained that, for a time, her great-grandmother rented out the house because she was working on another part of the island. "For those years, the family didn't get together because there was someone in there. But as soon as [the renters] went and bought someplace else, we were right back. And no one [lives] in the house, so that everyone [can] come."[110] In another family, all seven siblings share the taxes equally, and regularly clean the yard together, though no one lives there. These examples show how *'ohana* are working together to perpetuate communal extended family ownership and connections, within a Western property system geared toward privatization, partition, and sale.

Kupa 'ai au: Eating from the Land

> *Kupa 'ai au:* Native born, one long attached to a place through eating from the land[111]

> *I don't really come down here that often anymore, other than if I'm going to come fishing.*
> —Nani Paik Kuehu, Wanini elder and fisherwoman, 2015

Other strategies, such as food cultivation and gathering, transcend ownership, providing ways for families to maintain connection to areas where they no longer own land. In Hā'ena, for example, most area families have moved away but continue to do all of their fishing there. As *lawai'a* Keli'i Alapa'i explained: "Kalihiwai is much closer to where we live now, to [the town of] Kīlauea. But I don't go fish there because it is not our place; that is for the people from there. I always make the drive [thirty minutes] to come to Hā'ena, to fish in the same places where I learned from my *'ohana.*"[112] Fisher men and women express the importance of teaching their own children and grandchildren to fish in the areas where they learned from older family members.[113]

Connections to place continue to be rooted in food from that place. Families cherish the distinct taste of fish from their home reefs. One woman from Kalihikai described missing *nenue* (chub) from her home place because it had had a milder taste than the same species caught elsewhere. Her home reef does not have a particularly strong-smelling seaweed, *limu līpoa*, which she felt changed the taste of the fish that feed on that seaweed in other areas.[114]

Hā'ena *lawai'a* deliver the fish they catch to Hā'ena families who have moved to many different parts of Kaua'i, including Anahola, Kapa'a, Kīlauea, and Wainiha. One *lawaiʻa* regularly took fish all the way to the other side of the island, a two-hour drive, to Waimea, so that his uncle and family could eat fish from home. Families who have moved as far away as the West Coast of the United States also receive fish from Hāʻena for special occasions, such as high school graduations or first birthdays of grandchildren.[115] This sharing keeps Hāʻena families connected to their home, even though they can no longer live there. Here, *lawaiʻa* who return to their home areas to fish and gather for other community members fulfill a contemporary *konohiki* role. Though they are distributing fish far beyond historic *palena* (boundaries), they are maintaining social *palena* by keeping an entire community eating the same customary foods from their ancestral lands and ocean.

Ēwe, ʻŌwi: Birth to Bones, Maintaining Genealogical Ties

> *Ēwe hānau o ka ʻāina*
> Natives born of the land from generations back
> *I ke ēwe ʻāina o ke kūpuna*
> In the ancestors' family homeland[116]

> *This is our piko. This is where we come to rejuvenate and get away from the hustle and bustle of work. This is the one place you can come and reconnect. My grandmother's ashes are in this ocean here. Many of our family are here.*
> —Jamie Pānui-Shigeta, 2016

Ēwe, one word for "natives" of a place, also means afterbirth or placenta, attached to a baby's *piko* (belly button or source). Contemporary Hawaiian *ʻohana* continue to bury our children's *ēwe* or *ʻiewe* (placenta) in the soil of family lands as a means of connecting our offspring with their source. Other words for "native," such as *kulāiwi* (native land or homeland) and *ʻōiwi* (native) refer to those whose ancestors' *iwi* (bones) are planted in the land, perpetuating a cycle of generations rooted to the same soil. Families maintain genealogical ties to lands by returning to spend time with family members, both living and departed. Aunty Carol and Uncle Wendell Goo, both retired, drive to the coast where they grew up to walk early every morning. Though their families have moved away, walking allows them to check on the area and meet new community members. As Aunty Carol's sister says of her brother-in-law, "He knows all these new guys [who have moved here] 'cause he walks up and down."[117] Another couple in their sixties, Uncle Blondie and Aunty Nadine Woodward, grew up just houses away from one another at Kalihikai. Both of their families have had to sell their land at Kalihikai, and they now live half an hour away in Anahola, on Hawaiian Home Lands.

However, they make it a point to return to Kalihikai park for all family gatherings, holding their anniversary celebrations and grandchildren's birthday parties under the same tree where they were married, just across the street from the lands where they grew up.[118]

Public lands, such as county parks, provide means for families to gather in their home areas. Although parks are increasingly busy, they offer affordable space and facilities for reunions, as well as camping in three Halele'a communities. Some interviewees in their twenties and thirties explained that, as long as they can remember, their extended families have gathered to camp each year at the Kalihikai county park. Now they realize this is because family members of their grandparents' generation used to live in the area, and they have ancestors buried across the street from the campgrounds.

Though 'ohana may no longer own land on which to bury departed family members, they find other ways to return loved ones to their ancestral home, such as scattering ashes in the ocean across the street from former family lands. "A lot of my family, after they pass, we put their ashes right there. Right in the water [across from the house]. So we always feel safe.... That is our communing area. We need to talk story with them, we go swim. We ask them to send us the nice waves, help us fish. It's just part of our life cycle."[119]

Kalihiwai, Kalihikai, and Wanini 'ohana members at community research presentation at the Kalihikai beach park. (Photo by Ron Vave)

Hoʻokupa, Hoʻokamaʻaina: Perpetuating Knowledge of Place

Ua noho au a kupa i kou alo
I have stayed and become accustomed to your presence[120]

Our keiki [children] can hear about it from our kūpuna, but will they experience it for themselves?
—Olena Molina, 2016

Families work hard to perpetuate ties to the land by sharing knowledge and familiarity of home across islands, oceans, and generations. Though today's children are growing up far from ancestral homelands, adults find ways to pass on lessons of these places from the elders who raised them. One woman, kuʻualoha hoʻomanawanui, who frequently spent time with older family members in Wanini as a child, described how her great-aunt taught her daily and seasonal cycles by being still and observing:

> When I was little, three, four years old, every afternoon when the tide would go down, my grandma's oldest sister would go across the little street to the beach [with me], and we'd sit down. We never jumped right in the water. We'd sit there and she would just say, in the *kūpuna* way of not being very verbal, "The tide is going out." And we'd just sit there. And sure enough, after a while, when you grew up without iPads and cell phones, and all those distractions, you can literally see. . . . So till this day, I can look around the corner and go "well the tide's coming in, oh the tide's going out."[121]

This type of knowledge of place, said kuʻualoha, "isn't going to be on a map or in somebody else's *moʻolelo*. This is in their observation of being fisherpeople for generations down here." She went on to explain that she feels responsible for transmitting this knowledge shared with her to her nephew, the oldest child of the next generation, though he is growing up far away. "My nephew . . . lives in Virginia, so when we can, whenever he's home [in Hawaiʻi], we take him and pour knowledge into him."[122]

Kainani Kahaunaele set out what she wants her three children, growing up on the other side of the island chain, to know about her home:

> I want them to know what fish Kalihiwai is famous for. . . . Because that is what we ate often. And when the [certain fish come in], oh it was so good. I want them to know about the shark. I want them to know about the swinging rope. I want them to learn how to dive from that particular flat rock that we always used to dive from. I want to create a space, and a place in time, for the family to continue things that our family did. Because we can do that any place we go . . . ,

but it is more important to go to this beach. And even if we cannot live over there, and we create the experiences, I feel responsible [to do that].[123]

Younger generations also enhance their knowledge of home through formal studies. As Kamealoha Forrest of Hāʻena, who has become a scholar of Hawaiian language, explained of his work with archival resources, "Learning from the past, they kept such good records. Good enough that we can renew things, . . . follow those steps, and . . . teach. . . . That's what I see as my *kuleana*."[124]

Through sharing and applying their learning, young adults of Haleleʻa restore names found in centuries-old newspapers and chants to their home landscapes. Community members work to revitalize ancestral place-names such as Kahalahala instead of "Tourist Lumahaʻi," or Kauapeʻa instead of "Secret Beach," and discourage nicknames, such as "The Y" for Kalihiwai. The name Kauapeʻa means "rain bringing wind that fills the sails of the canoes." Kalihiwai means "the edge of fresh water." The richness of these names and the information behind them is lost with pat nicknames.

Community members also engage their creative talents to create *oli* (chants) and artwork, and share photos to honor their home: "It can help develop your appreciation for your home when you start focusing time on creating things about your place. Such as songs, such as portraits, such as sculptures. . . . Our home has inspired our cousins."[125]

Naming children also perpetuates *kuleana* and connection. As Kainaniokalihiwai (whose name means "the beautiful ocean of Kalihiwai") explained, "I'm just very grateful to be part of that place. My mother, because she is so fond of it, she named me for that place. . . . And so, more obligation for me. You know, I have to make sure I do something to live up to the name." Her younger cousin, Kaulana, in his early thirties, just named his first son, the newest baby in their *ʻohana*, Kalihiwai.

Kamaʻāina and *kupa ʻāina* relationships are built over generations of daily interaction with place. Many *kupa ʻāina* families now live and raise children far from their home areas and do not get to frequent them daily. Still, they try to transmit knowledge and responsibility through occasional visits and creative means such as hula, chant, art, song, names, storytelling, Facebook, and Instagram to keep faraway places close.[126]

Maka'āinana, Mālama: Attending and Caring for the Land

Hana ka lima, 'ai ka waha
Turn your hands down to the earth, your mouth eats
—Tom Hashimoto, Hā'ena[127]

It is that beach property, romantic fantasy that they are buying, because it is a beautiful place, good surf, and you know you can write home about it, like a beautiful picture. Whereas our families took care of that place as a matter of responsibility to feed the family, to learn about the area, to honor our kūpuna there. . . . So my wish for the future is to grow our children with that sense of responsibility being the forefront.
—Kainani Kahaunaele, 2015

The word *maka'āinana,* often translated as "commoners," literally means people that attend the land.[128] Many community members who have lost family lands find ways to continue taking care of the area. Kainani remembered how, as a child, older family members taught her and her cousins to care for their home area, connecting work, learning, and fun: "By then [her childhood] we had a big group of kids. And so it was the right time, since the family was bigger, [to] get back and start learning and making the connections: swimming, learning how to swim, learning how to surf, and how to take care of the sand, no trucks on the sand, park here, not there, and taking care of the yard, move this rock, not that one." Two of Kainani's cousins, twins in their late twenties, laughed that they miss doing the yard so they could go swimming. One looks down and in a quieter voice says, "Our younger cousins will never get to experience that, learning how to take care."[129]

Some community members said the hardest thing about losing family lands is not being able to care for them anymore. It is difficult to see former properties overgrown, because their parents and grandparents always kept the yard neat, expecting the same of their children. One eighty-seven-year-old woman from a Wanini family said she has to look away when she drives by the land where she grew up. Since Princeville Development Corporation took the property a few years before, the yard is not neat like her mother taught the family to keep it.[130] In two cases, family members have arranged with new property owners to do yards at no cost, so that they remain in the condition their elders would want. Like historic Hawaiian connections to land, the sense of responsibility among these families extends beyond the boundaries of ownership or distinct properties, beyond individuals or the present generation, to encompass caretaking of entire *ahupua'a.*

Hoa 'Āina: Collective Actions Bound to Protect Land

> *It's a place people go to, but to see it as its living being, you go there, you clean it, you take care of it, protect it from people that will do it harm. Like you would anybody, any little sister, little brother, older person. So that, that's what I think it means to really care for and see a place as family. That you would lay your life down for that place.*
> —Kamealoha Forrest, 2016

Hā'ena

Today, descendants of the original Hā'ena Hui members, who organized to collectively buy and hold the entire *ahupua'a* of Hā'ena through the 1960s, continue to care for former Hui lands. They have formed a contemporary Hui, open to any family who can trace their ancestry to Hā'ena prior to the Māhele, called Hui Maka'āinana o Makana (The Association of Maka'āinana of Mount Makana), after a prominent peak famous in many stories of Hā'ena. The Hui has negotiated a stewardship agreement with state government to restore and care for taro patches within the state park. Four Hui families farmed taro and lived on these lands until the state condemned them for a park in the mid-1960s.[131] Now members of these same families,

Hā'ena Hui members and supporters at the *lo'i* at Kē'ē 2016. *Front row, left to right:* Pat Schrader, Ua Hashimoto, Presley Wann, Mike Olanolan, Keli'i Alapa'i and Nick (Olanolan 'ohana). *Back row:* Mehana Vaughan, Pi'ina'e Vaughan, Jody Hashimoto Omo, Colleen Wann, Kawika Winter, Ted Blake, Wendall Olanolan, Butch Schrader, Samson Mahuiki, and Tom Hashimoto. (Photo by Kimberly Moa)

including the last farmer to grow taro here, are among the Hui members who gather two Saturday mornings each month to take care of the area. They work from morning till afternoon, then enjoy a potluck lunch together while holding their Hui meeting to discuss other community projects.

Hui members have cleared the *hau*[132] bush that overran the area, restored ten *lo'i*, and are slowly working toward restoring the entire ten-acre taro patch complex. These families hold fishing camps here, teaching the children of the community, including their own grandchildren, skills such as making and patching nets and catching, cleaning, and cooking particular fish. Families also hold reunions and camp at the taro patches, where the Hui maintains a small shelter for eating and two portable toilets.

Though the taro patch area is within the state park, public access is controlled by a gate with keys held by the Hui. In this way, the Hawaiian families of Hāʻena have negotiated access to a small area of their *ahupuaʻa* based on taking responsibility to care for it. Unlike other *ahupuaʻa*, where families have to gather in public county parks, the Hui's stewardship agreement provides a small area apart from the seventy-five thousand visitors who visit Hāʻena each year.[133] Though just off the highway, and in full view of the visitor traffic, the *lo'i* provide the community with a space of their own within the *ahupuaʻa*. One community member in his thirties captured the joy of being able to stay in Hāʻena overnight instead of driving from his home forty-five minutes away: "It was wonderful, to wake up in the morning with no one else there, . . . up early, no crowds, to see it like no one else gets to see it, the peacefulness, so beautiful."[134] He went on to describe fishing before tourists arrived at the beach, remembering one particular occasion where he and friends surrounded a school of *akule* early in the morning, then fed the entire community.

This *lo'i* area has become a *kīpuka* for Hāʻena families, most of whom can no longer live in Hāʻena, providing a place to build community and enact *kuleana* by regularly gathering to care for their home together. Their caretaking has served as catalyst for other community efforts, such as management of the fishery, by providing a place where community members can host government officials, conservation organizations, and other Hawai'i community groups. The immaculate *lo'i*, neatly mowed and weed-whacked, with its flourishing *kalo*, provides tangible visual evidence of the community's ability to work hard, work together, and effectively take care, thus positively influencing other collaborations that devolve management authority from the state.

Waipā

The 1,600-acre ahupua'a of Waipā, stretching from Mount Mamalahoa into Hanalei Bay, represents another important community *kīpuka*. The Waipā Project began in 1982, when a group of Hawaiian *kūpuna* from Halele'a, along with their families and supporters, organized to preserve the Waipā ahupua'a from proposed luxury resort development. After four years of negotiations, the landowner, Kamehameha Schools, agreed to lease the land to the group, under the name Hawaiian Farmers of Hanalei.[135] The *kūpuna* aimed to restore the ahupua'a so that it could support a subsistence lifestyle and serve as a *kīpuka* (place of shelter and restoration) for cultural practices.[136] Hawaiian Farmers restored historic Hawaiian irrigation systems in the valley, replanted *kalo* (taro) and began a weekly farmer's market. They began "poi day," bringing community volunteers together to hand mill the Hawaiian staple food every Thursday. To enhance access to healthy, traditional foods, Waipā poi is distributed to local families across the island of Kaua'i at low cost and gifted to elders for free. The group formed the nonprofit Waipā Foundation in 1994, with educational programs that integrate Hawaiian culture and environmental science in *'āina*-based activities.[137] Waipā's summer program, youth internships, and field trips for area schools serve thousands of learners each year and now employ former students as teachers.[138] Through tireless grant writing and fund-raising, Waipā has become one of the largest area employers of local youth, who otherwise work in resort or visitor service industries such as restaurants, clothing stores, landscaping, and cleaning. Most young community leaders have been mentored through Waipā programs. In recent years, Waipā has piloted limited programs for tourists in order to enhance financial self-sufficiency. These include farm-to-table tours, dinners focused on consumption of invasive species, and participation in restoration and workdays as part of an authentic local experience.[139] Waipā also provides a quiet place for area families to camp, gather, learn, and work together, and serves as a gathering place and center of the community. For over thirty years, Waipā has helped community members get to know, care for, and cultivate the *'āina*, while providing food and nourishment.[140]

Both the *ahupua'a* of Waipā and the *lo'i kalo* at Hā'ena provide sources of learning and connection, not only for the families descended from these places, but also for the broader Kaua'i community. Everyone is welcome at monthly community workdays. Friends and supporters, of many different ethnic groups, come regularly. For these families, such as my own, community *kīpuka* provide places to teach our children practices of our ancestors

Uncle Charlie Pereira (left) and Aunty Audrey Loo (right) clean *kalo* to make poi at Waipā. (Photo by Megan Juran)

Aunty Annie Hashimoto (left) and Aunty Honey-Girl Hoʻomanawanui making poi at Waipā. (Photo by Megan Juran)

that might otherwise not have continued in our own *ʻohana*, to connect with other families of our community in a deeper way, and to give back to places we love through working together. Both sites regularly host groups of visitors from around the world who also come to work and learn. These *kīpuka*, and Haleleʻa and Koʻolau families' efforts to care for them, help people reconnect and take on responsibility for their own home areas. As one interviewee explained, "There is a way that you can find more appreciation, not just for [this place], but also for your own areas. That is the goal."[141]

Kīpuka: Reserves of Replenishment

As change continues, at least connection will remain the same.
—Kainani Kahaunaele, 2015

With all that encroachment, you still find peace.
—Blue Kinney, 2015

Hiking across a recent lava flow, there is no sound but wind sweeping rock and the monotonous crunching of steps, shoes slowly filling with dust. Heat rises from the expanse of black stretching in all directions. There is neither shade nor shelter, except crouched in a hole or swale in the cooled magma. Occasional birds fly swiftly past, headed elsewhere.

Kīpuka, where a lava flow has split and spared an area of the forest, are felt before they are even seen. Sometimes *kīpuka* are tucked away in a hidden crater. The air cools, wind softens, and native honeycreepers flitter out from the perimeter of trees. The forest floor is damp with moss, and each blossom on laden branches will give way to pods that expel millions of seeds like dust. *Kīpuka* are more than refugia, islands of what the landscape once was. *Kīpuka* provide reserves to restore the essential character of surrounding areas, to reseed communities.

Besides oases of forest, the word *kīpuka* describes any variation or change of form in the landscape: a calm place in a high sea, a deep place in a shoal, or an opening in a forest or cloud formation. Each of these changes provides a means by which to see, understand, and navigate the continuous stretches that surround. *Kīpuka* also means a poncho or shoulder cape, crafted to provide protection from wind and rain, as small parcels of land within a changed landscape provide a place of shelter. Lastly, *kīpuka* means loop, lasso, or snare, a tool to capture something and bring it toward you, as *kīpuka* provide a way to call families back to the land, calling on connections that, though hidden, still endure.

Investigating *kuleana* in relationship to land expands the meaning of *kuleana* to include maintaining ancestral ties, perpetuating family practices and presence, cultivating and eating from the land, learning and teaching about it, and organizing to care for and protect places collectively. Stories of Halele‘a families' responses to shifts in the landscape and community of their home teach the importance of continuing to fulfill *kuleana* when ownership is lost, even as access becomes more difficult.

Today, Hawaiian families of Halele‘a are experiencing what their forebears endured on a national level after the Māhele and illegal overthrow of the Hawaiian Kingdom—the expansion of foreign land ownership and wrenching loss of access to places they have tended and used for generations. *Kuleana* lands awarded after the Māhele failed to encompass the full relationships of *maka'āinana* to their *ahupua'a*, just as current-day loss of the remaining small parcels does not terminate broader landscapes of connection and responsibility. Though struggling from and, in many cases, saddened by the loss of individual properties, Halele'a families express enduring connections to the broader landscape and to one another. Their stories and voices honor their homes, whether those homes remain physical places or memories. Their voices express *aloha 'āina* (love of land) and *kuleana*, which transcend ownership and, unlike property, cannot be taken away, gated, or sold.

'Āina needs people to connect, call out, and care for it in order to retain its character, just as families need *'āina* to retain who they are. A place and its people are one and the same. Relationships to land are inseparable from the exercise of *kuleana*. People of a place have responsibilities that take work: perpetuating stories and names, watching for changes and chronicling them, protecting, welcoming, learning, and teaching. These responsibilities are larger than any one piece of property. Because beaches are public, though people cannot access former house sites, they can return to surrounding coastal areas. Families can continue to select final resting places in the ocean or cemeteries of their home area, ensuring that, like *ea*, life of the land, quiet presence endures. Also, because connections to the land were communal and social, families that stay connected to one another stay connected to their land, even if they cannot go there. When they do, the land sings with joy.

As landscapes have transformed from farming and fishing communities to luxury vacation retreats, Hawaiian families who can no longer live on the coasts of Ko'olau and Halele'a visit when they have time off from work. These returns home happen on weekends or holidays, when visitor traffic too is at its height, lumping *noho papa* with streams of newcomers and tourists. Yet, even within the context of beach-going and recreation, families seek to exercise responsibility to care for places and other users. Maintaining homes for these families along every stretch of coast would benefit both the place and its visitors by returning eyes to the land and providing hosts, enhancing safety, protecting resource health, and restoring ancestral character of *'āina*.

Even very small parcels can serve as *kīpuka* where families return to teach, learn, and enact *kuleana* to the land and to one another. Models of land held in common (with the help of land trusts or nonprofits), rather than by one family or owner, could lower property taxes, reduce financial burdens, and allow families to share land with other local families who have lost their *'āina*. Families could contribute in many ways that are not solely financial, offering labor to upkeep houses and yards, bookkeeping help, event organization, teaching of cultural practices, and other forms of caretaking rather than capital. Benefits would again be collective, rather than private.

Kīpuka in each *ahupuaʻa* would serve as generative spaces where area families and other *hoa ʻāina* share in work, learning and caretaking without ownership rights, reinstating relationships to land based on *kuleana*, and restoring reciprocity of rights and responsibilities. A network of lands held as cultural *kīpuka* within every *ahupuaʻa* could also provide places for controlled numbers of visitors to not only learn about but also feel and contribute to caring for the essential character of each *ahupuaʻa*. *Kīpuka* provide seeds of cultural knowledge and practice to spread to other places where these practices are discontinued. These lands also provide areas for farming and fishing, restoring sections of *ahupuaʻa* as working and feeding landscapes versus solely places of recreation. While strengthening ancestral ties of *kupa ʻāina* families to one another and the land through regular gatherings, *kīpuka* also provide space to build new connections and futures grounded in the daily exercise of *kuleana*.

Moʻolelo: Wāwā's Legacy

Every time I visited Kalihiwai Bay to swim, I was saddened by the emptiness of my friend Kainani's family home—its peeling paint, sagging steps, and overgrown yard. Our whole community had been used to seeing their *'ohana* there; playing cards on the porch, reminding people not to drive their trucks on the beach, weed-whacking, calling out to us as we slowed down to wave out the windows of our cars, inviting us to come eat. Without their family's presence, I would leave visits to this bay, my place of refuge, feeling raw rather than restored. I started crossing the street to stand in the lonely yard. I addressed Tūtū Makaleka, her daughter Wāwā, and the ancestors of this *'āina*, asking that they show us a path to get the property back for their family.

Six months after Tūtū Makaleka Day, Kainani, a small group of her family members, and I returned to the yard for a meeting. I had identified the new owners, partners in a Hawaiʻi realty firm who had purchased through forced auction. I invited them, along with a dear friend engaged in land trust work on the island, to meet with family members at the house. I had no clear plan except that the owners should hear the family's stories of this place, their connection to it, and how they lost it against their will.

One owner of the firm confessed to the family that he had never even set foot on the property. Though he and his wife lived nearby and walk the beach across the street regularly, he said it just didn't feel right. His partner, who actually attended the auction to bid, saw all the family members in the courtroom. He said he was worried they would be angry and confront him. Instead, after he won the bid, they gave him hugs. He said that growing up with parents who had moved to Kauaʻi, he had never experienced anything like Kainani's families' extended ties to this land and to each other.

At our meeting with the new owners, family members told of spending time together at the home as kids. They slept twenty cousins shoulder-to-shoulder on the floor of the living room, and bathed all at once in a tub in the yard with the cold hose. They played on the beach all day until their Aunty Merilee's whistle called them home for lunch and dinner. Food was simple and rules were strict.

When these children grew older and started their own families, they kept a calendar. Every *'ohana* took a month to use the house and care for the yard. No matter which branch of the family was on the schedule, everyone would come, bring their children, and spend time together. Family members express that they are close today because of this place. As one of Kainani's aunties explained, "This is our *piko* [center] place. We start our lives here, learn and grow, and when it is time to go home, this is where we want to return to, where we want to be."[1]

Another of Kainani's aunties described the pain of the auction process. "We tried everything we could, and then we had to let it go. We had to say 'OK, we'll still go [to Kalihiwai], we'll go. We just have to go on the other side of the road.' And our children, even though they didn't experience it like us, this is their home."[2] Family members at the meeting expressed their gratitude to the new owners for the chance to set foot on the property again, something they never thought would happen. As one aunty put it, "Whatever comes of this, there is so much emotion to be back here, just being back on this side of the road after two years. Whatever happens, today is a good start. There is no rush to decide anything. We will all now go back and talk to our families. But whatever is meant to be will be."[3]

Together, our group of family members, temporary "owners," and community supporters have continued quietly working on ways to return the home to the family. We are trying to develop a *kuleana*-based model that can be used to help other area families in similar situations. After our meeting, the realty firm took the property off the market. My land trust friend approached a generous community benefactor for help and invited this woman too to meet with Kainani's family. The woman and her husband, who started a successful dot-com business, moved to the area fifteen years ago. They support multiple community causes and provide a trail to maintain public access through their extensive land holdings. Moved by the family's story, the woman offered a property of her own to trade for Kainani's family's parcel. Though the second property was substantially lower in value, all of the partners in the firm agreed to the trade.

This generous woman sees herself as temporarily holding the Kalihiwai property for Kainani's family, as they create a plan for the future. They have started a nonprofit to lease the land, with the goal of eventually buying it back. The *'ohana* agrees that no relative will live there, but all will be able to stay, using a calendar again to take turns. They would like to include times for other families who have lost lands in the area, to hold their own reunions. They also want to have fishing camps, *haku mele* (song composition workshops), and educational programs for area families and youth, including those recovering from drug addiction. They plan to repair the house just as it used to be, and are working with the current

Descendants of Tutu Makaleka gather to celebrate and bless their return to the family home on November 25, 2017. This first gathering is intended to become an annual Thanksgiving weekend tradition. (Photo by Kilipaki Vaughan)

Descendants of Tūtū Makaleka's granddaughter, Kanani Pānui Kahaunaele, gather on the newly rebuilt porch steps around her husband, Papa David Kahaunaele. Each branch of the family posed for a picture in their "Wāwā's Legacy" T-shirts to commemorate the happy day and also show how each person is connected. The family asked that community members who helped work on the return of the property pose for a photo too. (Photo by author, November 25, 2017)

owner to create easements that protect the land from further development in perpetuity.

On the Saturday after Thanksgiving, November 2017, the family held a celebration to launch their nonprofit and to bless their return home. They named their nonprofit "Wāwā's Legacy." Over two hundred family members gathered, representing many but not all of Wāwā's forty grandchildren, eighty great- grandchildren, and seventy-two great-great-grandchildren, along with descendants of her other siblings and their mom, Tūtū Makaleka. The family also invited other Kalihiwai families and those who had helped get the property returned, including the current owner, who shared how honored she was to be included. Children of all ages were everywhere, throwing a football behind the house, drawing on the butcher paper stretched over tables under two easy corner tents, sitting on their grandmother's laps, and handing out folded programs for the day's blessing.

Tables overflowed with food and desserts that family members prepared. Tūtū Makaleka's quilt was placed with a display of family photos on the porch, alongside sign-up sheets to assist with repairing the house. A sudden deluge of rain brought everyone scurrying under the tent just in time to hear recollections of Wāwā by her children and grandchildren and to sing songs that she had taught them. Family members stayed all day and into the night catching up. The family plans to make the celebration an annual event, as Kainani says, "the beginning of much good work to come."

NOTES

1   Jamie Pānui-Shigeta, personal communication, 2016.
2   Nani Pānui Sadora, personal communication, 2016.
3   Ibid.

Moʻolelo: E Alu Pū (To Come Together)

October 2, 2014: Preparing for Public Hearing on
Hāʻena Community Fishery Rules

Like most community meetings in Haʻena, this one is held at the *loʻi* (taro patch). Some chairs and a bench form a makeshift circle in a tin-roofed, forest-green plywood shed. It is open on the three sides that don't face the prevailing trade winds. Discarded hotel pool chairs are lined up neatly for extra seating. Boards to pound poi are covered with tarps, roped secure. The sand and dirt floor is raked.

A fire burns just outside the shed, wood smoke meandering occasionally into the gathered crowd of 150 people. The first *loʻi* terrace begins three feet from the shed. One could weed the taro while still within earshot of the discussion. The many people sitting just beyond the shed's roofline can gaze on

The *loʻi* at Kēʻē serve as a gathering place cared for by Hāʻena families. (Photo by Anuhea Taniguchi)

Members of E Alu Pū from across Hawaiʻi gathered at the Kēʻē *loʻi* to support Hāʻena. (Photo by Kimberly Moa)

the entire height of Mount Makana, green-covered rock faces lit gold by the sun, *koaʻe* birds soaring on the occasional updraft. This same wind, Unupali, carries firebrands thrown from the cliff at night, spiraling out over the *loʻi* to the sheltered bay of Kēʻē, in a ceremony called *ʻōahi*.[1]

Inside the shed, flip-chart paper lines the only wall and papers the old glass-front restaurant fridge now used for storage. One paper has large color photos of the governor and key legislators who Hāʻena will need to support their community-based fishery rules. Another paper offers pointers for presenting testimony: "What's your ONE message? You have THREE minutes." The group of Hāʻena community members, state agency personnel, conservation NGO staff, community advocates, and fisherpeople from twenty different Hawaiʻi communities have been meeting all afternoon to prepare for tomorrow's public hearing, the culminating event in the formal review process to determine whether Hāʻena's community fishing rules become law. Many in the group have never been to Hāʻena before, though they have supported Hāʻena's efforts through email, meetings, and lobbying for years. Collectively known as E Alu Pū (to gather together like a school of fish), these community leaders who engage in local governance of land and waters across Hawaiʻi gather twice a year to visit one another's communities and learn from each other's work. Everyone has come to testify in support of Hāʻena's rules, the first community-developed fishing rules to go before the state for approval to become law. Those gathered represent communities that have helped pave the way and those that aim to follow.

The group has just been coached on offering effective testimony from a policy and advocacy expert hired by the nonprofit Kua 'Āina Ulu 'Auamo, or KUA (backbone),[2] which facilitates E Alu Pū and has organized this training. Everyone has been split into groups of ten people, to share in a circle what we'll say in our testimonies. Students in each group, including seven of my own, sit with laptops at the ready to type notes. Tomorrow, these notes will be returned in printed form so each person can practice. In my group, no one is sticking to the rules or three minutes. After ten, we are only on the first person.

Uncle Samson Mahuiki sharing stories.

Uncle Samson Mahuiki is full-blooded Hawaiian and one of the oldest *kupuna* of the community. He is sharing stories of the Hawaiian horses he trained and rode solo at night down the narrow eleven-mile Kalalau Trail along the remote coastline arcing west from here. These days, highly trained teams of Kaua'i County fire specialists helicopter in to rescue hikers. They saved sixty tourists on a single day in the summer of 2014, when the river flooded and closed the trail. One woman was swept to sea when she tried to cross the raging waters, along with her husband who tried to save her. Both drowned. In the 1960s and 1970s, when ten to twenty tourists hiked the trail per day, instead of three hundred, search and rescue consisted of Uncle Sam and his horse. No one interrupts or glances at their watch, just listens, entranced. We have eaten lunch of fried fish caught that morning from the reef out front, sweet poi from the taro patches spread around us, rice and beef stew, all prepared by Hā'ena Hui members. There is risk that with such *mā'ana* (pleasantly satisfied) stomachs, everyone will be asleep before we get halfway around the circle. Uncle Kealoha, who trained his horses on the beach until ten years ago and still makes their saddles, leans against a nearby java plum tree, quietly singing and plucking his guitar.

This morning, two Hā'ena community leaders toured the group along Hā'ena's coast. Uncle Tom Hashimoto, Hā'ena's oldest fisherman, shared some of the place-names for fishing holes taught to him by his father. He took them to Mākua, the nursery lagoon the community has decided to close, not only to fishing but to all human activity, in order to help fish stocks replenish. Uncle Presley, the current president of Hui Maka'āinana o Makana, is a shy man of few words. He took the group to see his 'ohana's *kuleana* land, which once stretched from the highway to the sea. What remains is named Kamealoha (that which is beloved), five hundred square feet of former taro patches and a shed, wedged in between portions of the land sold to Eagles guitarist Glen Frey and movie star Julia Roberts. The patches no longer receive water, as the irrigation ditch has been dammed up to form a koi pond on one of the purchased parcels.

Though her siblings sold their lots, Uncle Presley's mother, a quilter, continued to sell Hawaiian quilts long after her arthritis made it excruciating to sew, to afford the taxes to keep her portion of the land. She carried on for her grandfather, Hailama, a gifted waterman who kept his canoe just out front of their *kuleana* land, sailing regularly to Ni'ihau and to Nihoa, the start of the remote Northwestern Hawaiian Islands. In the 1930s, Hailama moved to Honolulu to get a wage-earning job so he could pay taxes. He was homesick, and eventually died and was buried in an urban cemetery. One day Uncle Presley hopes to bring his great-grandfather home and bury him here at Kamealoha. Uncle Presley also scat-

tered his mother's ashes here because she loved this place, though she could rarely afford the time and money to come home from Honolulu. Uncle Presley does not know how much longer he can afford to pay the taxes himself.[3]

After retiring from forty years as a heavy equipment operator, Uncle Presley planned to work on restoring the taro patches on the land. Instead, he was elected to lead the Hui, taking over from Uncle Sam's son, Keli'i Alapa'i, who succeeded Jeff Chandler, and has been too busy. This morning, from atop the dunes where bones of his ancestors sometimes peek out after high surf, Uncle Presley pointed out his fishing grounds for the group. "From that reef to that reef. Right here. I don't go anywhere else." One woman asked how the fishing has been lately. Uncle Presley, looked down then into her eyes, "I couldn't tell you. I haven't been fishing since I got involved in this whole rules thing five years ago. It just doesn't feel right."

Uncle Presley Wann and Uncle Blue Kinney doing dishes at a community gathering. (Photo by Anuhea Taniguchi)

October 3, 2014: 5:00 A.M.

That night, Uncle Presley dreams the same dream he has been having all week. He has caught too many fish and is driving from house to house trying to give them away before they spoil. He awakes unsettled to find that the zippers on his tent, which he'd shut tight when he went to sleep, are open. A chill wind is blowing through. He figures his mother is trying to tell him she's around, to keep him alert for something.

At sunrise, the group starts to clean one of the taro patches. Students, scientists, nonprofit directors, public relations specialists, fisher men and women, farmers, grandmothers, *keiki* (children), wade into the knee-high weeds and sink into the mud. They work in a flowing line, pulling, shaking clods of dirt from roots. They hand the weeds one at a time toward the banks, where others pile them, then carry the heaps to bushes at the edge of the patches. The old-timers' T-shirts are spotless despite mud wrapping up to their calves. Occasionally, one of the younger ones at the end of the line flings weeds to the bank without shaking sufficiently. Mud scatters over folks nearby, causing exclamations, laughter, and teasing. Otherwise, the group works quietly, steady, sun cresting the ridge and bathing the spaces between them in flickers of light.

Members of E Alu Pū work in the *loʻi* the morning of the hearing.
(Photo by Emily Cadiz)

When the *lo'i* is entirely weeded, the group stomps the entire patch smooth beneath a layer of six-inch-deep water. A few stragglers step down the last hillocks of mud, while the group helps each other scrub in the hose spraying fresh water from the stream. They wash off the taro patch mud for the next kind of work, a practice public hearing to be held over lunch. Everyone is handed a printed testimony based on the notes typed while they talked yesterday. Each person stands to practice speaking in the same order they will speak at the hearing. As the oldest living *lawai'a*, Uncle Tom Hashimoto will testify first with his family, followed by all of the Hā'ena families in turn, then those from other islands. The plan is to get to the hearing early and sign everyone up to testify, filling the roster with supporters.

My baby starts to fuss. I take her out of the crowd and walk behind the line of sugarcane, thinking I can put her to sleep. I'm surprised to see Uncle Jeff sitting on the ground near a rock, lighting a cigarette. "Hey," he says, as if it hasn't been years. His skin is pockmarked and sallow, but he looks straight at me with that same fierce stare, lifts his chin. "How's your mom and the family?" I bend down and try to give him a hug, but he just sits there stiff. I introduce the baby, and tell him I've had two *keiki* since I last saw him, to which he nods. Then, "Uncle Jeff, I'm so happy you are here. This couldn't be happening without you. Come see everybody." He shrugs, stubs out his cigarette, and follows me slowly back toward the shed where his brother and sister are taking their turn addressing the crowd.

Aunty Lahela sees him and pauses, her eyes filling, "Come, brother." I walk him forward to join them. Uncle Jeff shuffles as if each step is an effort. Uncle Moku puts his hand on his back, and Aunty Lahela says, "These are my brothers, Moku and Jeff. Some of you may not know Jeff, but he's the one who got us into this. It is so good to see you brother." Uncle Jeff stands, thumbs slung through a black fanny pack, weight on one foot, looking beyond the crowd. He has the weathered wild look of someone who is not sure where he will sleep tonight. Those new to this effort do not know how changed Uncle Jeff is, but those of us who have been in this longer, who Uncle Jeff roped in, inspired, tasked, meet one anothers' eyes and find tears.

I wonder what Uncle Jeff would think of the rules now, these rules we are practicing to testify in support of tonight. These rules we will ask members of the State Board of Land and Natural Resources, none of whom will even be at the hearing, to vote on in yet another meeting and, after reviewing all of our testimonies, to approve. Uncle Jeff, who always told government representatives, "You are welcome here, in our place. But let's be clear. We are not partnering with you, you are here to partner with us." Uncle Jeff, who started this effort because he was tired of seeing tourists—trampling the reef, snorkeling up to him and spooking the fish just before he threw his net—arriving in ever-greater numbers. Uncle Jeff, who said, "It's the fishermen who are becoming an endangered species."

In my mind, I see him bounding across the reef, following his net after he threw, adroitly missing holes in the coral. I remember him striding along the shore, stopping every few steps to pick something up, smell a leaf, bite off a piece of seaweed, explaining how every bit strewn on the high-water line holds a clue to what is happening in the ecosystem. The rules seem so different now from the early brainstorming meetings Uncle Jeff led, with so much negotiated out to make everyone happy. Uncle Jeff was never about making anyone happy. He was about doing what he felt was right. For him the rules were just a step in the bigger picture, the community taking control of taking care of their home. If he could read and understand the rules now, would he still see them as a step in that direction?

Uncle Jeff leaves again within the hour. He does not show up for the hearing, which may be good, as he would scare people. Even when he was healthy, we never knew what Uncle Jeff would say in a public meeting, or when his voice would gather momentum and force till people felt yelled at,

Uncle Moku Chandler and Aunty Lahela Chandler Correa testify in support of the Hāʻena community fishing rules before the Board of Land and Natural Resources in Honolulu, on behalf of their ʻohana, including their brother Jeff. (Photo by Kimberly Moa)

"This is OUR place that you are supposed to be taking care of. But you [the state] are not doing your job. So just let us do 'em." Before ambling back through the bushes and starting up the highway, he turns and calls softly, "Hey, I gotta talk to you sometime, about a hula.... I need your help to make a throw net hula, capture the motions, it's a dance, a dance."

NOTES

1   Young men of this area are again learning, like their grandfathers, to scale the perilous cliffs on a foot-wide trail, carrying ten-foot hollow logs. They build a fire on the peak, light both ends of the logs, and feed them out into the wind. Without the right breeze, the logs plummet straight down the mountain, and once, in 1910, set the entire cliff ablaze. But the Unupali catches them and carries them dancing and spiraling out to sea. Sailing canoes once lined up in the bay, bearing young chiefs poised to catch a burning brand and touch the fire to their skin as proof of their steadfast love for some beguiling chiefess. Hāʻena is one of only two places in Hawaiʻi where this ceremony is performed for special occasions such as the visit of an *aliʻi* or other special guests, or to commemorate Makahiki season. These flying fires of Makana have blazed just five times in the past 125 years, with a lapse of ninety years between 1915 and revival by Uncle Moku Chandler in 2007, with his nephews and son. Uncle Moku, a man of few words, says simply, "I feel so lucky to have this place, and to keep this going so other people can experience it too."
2   Short for Kuaʻāina Ulu Auamo (grassroots people taking up the sacred *kuleana* to better Hawaiʻi). KUA's mission is to "empower communities to improve their quality of life through caring for their biocultural (natural and cultural) heritage." Their vision is *ʻāina momona* (http://kuahawaii.org/).
3   Their family cannot qualify for the *kuleana* tax break awarded to those who can trace their genealogy to the original awardee of a *kuleana* after the Māhele because of missing paperwork. Families in Hāʻena were born and died at home, without birth or death certificates for documentation.

CHAPTER 6

Kiaʻi

Carrying *Kuleana* into Governance

We gotta get back to the konohiki system, and maybe the konohiki is gonna be the community.
—David Sproat, Kalihiwai, 2015

You gotta believe in it and you gotta live it. If we're gonna make these rules then we gotta live it.
—Chipper Wichman, Hāʻena, 2011

This isn't about extra agencies being needed or extra enforcement; all we need is the ability to do what we know how to do, in a place [the families of Hāʻena] know best.
—Makaʻala Kaʻaumoana, Hanalei, 2014

Historically in Hawaiʻi, the people of an *ahupuaʻa* served as *kiaʻi*, guardians or caretakers of local resources, from fishponds to streams, mountain forests to coral reefs.[1] Though *konohiki* shifted, *makaʻāinana* families stayed in and watched over the *ahupuaʻa* of their ancestors across generations.[2] Under the territorial and, later, state governments, decision-making about natural resources shifted from local *konohiki* and *makaʻāinana* families to centralized state agencies. This chapter investigates how the people of Hāʻena are perpetuating their role as *kiaʻi* of their *ahupuaʻa* by working to restore local-level governance. Governance "refers to who participates in decision-making, the procedures and rules by which decisions are made and consensus is reached."[3] As they carry *kuleana* into governance, the families of Hāʻena are strengthening their community and state policy. For over twenty years, Hāʻena community leaders have developed collaborations with state government agencies. Over the last decade, the community created local-level fishing rules based on ancestral practices of the area, which have now become state law. The Hāʻena community, led by the Hui Makaʻāinana o Makana, is a model for over twenty other Hawaiʻi communities seeking to reestablish local *ahupuaʻa*-based fisheries governance.[4]

Hā'ena's coast, including the Kē'ē *lo'i*, circa 1924. (Hawai'i State Archives, Photo by 11th Photo Section Air Services USA)

Carrying *kuleana* into governance requires caring for *'āina* and community at home while also partnering to shape policy efforts at the county, state, and federal levels. Engaging in governance requires very different skills than fishing, including community organizing and outreach, public speaking, political strategizing, drafting legislation, networking, and using social media. Communities like Hā'ena work within existing systems of state government, while creating new institutions that enhance capacity for local governance.[5] This chapter begins with background on Hā'ena's rule-making effort, then shares the challenges of partnering with state agencies, Hā'ena's exhaustive processes for developing collaborative management, and disappointments with the outcomes of the rules, along with key achievements and lessons.

Background of the Hā'ena Fisheries Rule-Making Process

> *I limit myself because I see what it was like before. There were plenty fish! Not like today, you strain your eyes looking. Big like this tent, the pile of moi, and some bigger, the ulua behind, riding the wave, silver all in the wave.*
> —Tommy Hashimoto, elder Hā'ena fisherman, 2009

With formal management authority centralized in state agencies, fishing communities across Hawai'i have observed declines in their nearshore fisheries.[6] One state agency, the Department of Natural Resources and

Environmental Management, oversees 1.3 million acres of state lands, beaches, and coastal waters and 750 miles of coastline.[7] State fishing regulations are largely uniform across the main Hawaiian Islands, despite evidence that species spawn at variable times in different locations.[8] State regulations, including gear restrictions, catch limits, and seasonal closures, need to be refined to reflect ecosystem complexity. However, state agencies struggle to enforce even existing rules because of lack of funding and personnel.[9]

In 1994, in response to shortcomings in state governance, rural subsistence fishing communities across the Hawaiian island chain advocated for legislation to designate community-based subsistence fishing areas (CBSFAs) "for the purpose of reaffirming and protecting fishing practices customarily and traditionally exercised for purposes of Native Hawaiian subsistence, culture and religion."[10] CBSFAs allow subsistence harvest governed by community-generated rules and management reflecting local ancestral practices and values.

Communities across Hawai'i saw CBSFA designation as an eagerly awaited tool to revive culturally grounded, local-level management.[11] However, state agency personnel, tasked with implementing the law, viewed the legislature's directive as an inconvenience and, in some cases, a threat.[12] Over the next twelve years, the Hawai'i Department of Land and Natural Resources (DLNR), which "manages, administers and exercises control over the State's natural resources," did not designate a single CBSFA. A successful two-year (1995–1997) pilot CBSFA management effort at Mo'omomi on the island of Moloka'i resulted in increased fish biomass and abundance.[13] However, DLNR and its subsidiary, the Division of Aquatic Resources (DAR), failed to promulgate administrative rules for permanent designation of Mo'omomi after the pilot ended. DAR lacked funding, motivation, and expertise among their staff of fisheries biologists to devolve governance and implement collaborative management.[14]

In 2006, a group of Hā'ena community leaders with extensive experience working with state government decided to take a more active approach, rather than wait for designation by DLNR.[15] Instead, they worked with Kaua'i's genteel, Native Hawaiian state representative, Ezra Kanoho, to pursue direct designation by Hawai'i's legislature. Kanoho introduced Act 241 to declare Hā'ena a CBSFA in 2006, his final session before retirement. His fellow representatives voted for the bill out of respect for the senior statesman despite opposition from commercial fishermen and tour operators on other islands. The following legislative session, ten other Hawai'i communities, including three entire islands (Moloka'i, Lana'i, and Ni'ihau), also introduced bills for designation as CBSFAs.[16] None passed. Legislators cited the need to see evidence of Hāʻenaʻs success before designating more CBSFAs. As the head of the Division of Aquatic Resources at the time explained, "Hāʻena is important because they are going to set the precedent for how [comanagement of inshore fisheries] might happen in the future [in Hawaiʻi]. If itʻs a complete mess, [DLNR] is [not] going to go down this route again anytime soon. But if it works out, then you might actually see this trend toward gradual re-empowerment of communities."[17]

Hāʻenaʻs Act 241 required DLNR to work with Hāʻena residents to create and enforce regulations based on traditional coastal use. After passage of the bill, community members started an informal fisheries committee of ten to fifteen individuals, ages sixteen to seventy. The committee began meeting regularly with facilitation from a nonprofit conservation group, Hawaiʻi Community Stewardship Network (HCSN), now KUA, which helped draft the legislation.[18] To inform understanding of ancestral and current fishing practices, changes in resource health, and recommendations for future management, community members conducted twenty interviews with Hāʻena fisher men and women in 2006, adding to fifteen elder interviews already conducted in 2003.[19] The committee drew on these interviews to guide management planning. They identified species of concern and key threats, brainstormed rules to address both, and sought broader community input.[20] While developing rules, members of the fisheries committee also conducted three separate studies of coastal use assisted by university students.[21] Studies included human use counts, a catch per unit effort study to assess current levels of fishing, and a Manta tow survey, in which community members assessed substrate health along the coast.

The Division of Aquatic Resources required that proposed rules be stricter than existing state and federal regulations, constitutional, and easy to obey and enforce. Through ongoing negotiation with DAR, facilitated by HCSN, the committee ultimately settled on twelve draft rules. They gathered feedback on these rules through meetings with Hāʻenaʻs *kupa ʻāina* fishing families,[22] other coastal users such as surfers and commercial kayak operators, and the neighborhood association encompassing all Haleleʻa area residents. After five years of feedback, revisions, and translating the proposed rules into legal language, the Hāʻena community submitted their rules and management plan for approval to DAR in June of 2011. Both DAR and DLNR failed to respond to the community for over a year, then took two additional years to review the rules before scheduling them for public hearing in 2014. The rules were finally signed into law by Hawaiʻiʻs governor in August of 2015 after nine years, seventy meetings, fifteen rule drafts, and three public hearings in which 99 percent of testimony was in support.[23]

Challenges in Choosing to Partner with State Agencies

Ea: Fumes

This is where it gets cloudy in Hawai'i. . . . Because of the long history of everything, people don't necessarily want the government to be involved, but now they kind of feel like they have to, maybe.
—Former head, Division of Aquatic Resources, 2011

The choice to partner with state government agencies to restore local-level fisheries governance held challenges for the Hā'ena community. While some community leaders felt collaboration with state agencies was necessary to protect area resources and enhance local governance, others were more cautious. Key challenges included concerns regarding legitimacy of government regulation, risk of undermining informal community efforts, monopoly of community time, and bureaucratic delays.

Concerns Regarding Legitimacy of Government Regulation

Ea: Independence

People really have lost so much. Anytime you're talking about a culture where so much has been taken away . . . the more that stuff comes up.
—Debbie Gowensmith, Hā'ena rule-making process facilitator, 2009

In 1893, a small group of American businessmen backed by the US Navy illegally overthrew Hawai'i's monarchy.[24] Seven years later, Hawai'i was annexed without treaty or vote of the populace, in violation of international law.[25] Because of this history, influential community members throughout Hawai'i distrust and oppose collaboration with state or federal agencies viewed as representing an illegitimate, occupying government.[26] On the north shore of Kaua'i, people's sense of loss is fueled by the continued loss of family lands to wealthy Americans, increasing visitors recreating in rural areas, and shrinking beach and mountain access. In Hā'ena, some fishermen viewed even community-generated rules as more government regulation of their lifestyle and ability to feed their families.[27] One fisherman in his forties articulated concern for the cross-generational effects of area closures.

You close [the fishery] for, say, five years, . . . my boy is going to be gone [off to school], and through his period of time he could fish there. He will not know how to fish there. . . . One drought of the cycle of us, the human life. He will not learn . . . because he already will be in these distracting years, and he won't wanna be around fishing, he'll be doing other stuff."[28]

Debbie Gowensmith, Kevin Chang, and Shae Kamakaʻala of KUA handing out T-shirts at the *loʻi* before the hearing. (Photo by Kimberly Moa)

For rural Hawaiʻi families, fishing and hunting provide security, independence, and pride. Environmental policies to protect resources can be perceived—or branded by opponents—as threats to self-sufficiency, Native rights, and local lifestyles, provoking opposition to local fisheries management efforts throughout Hawaiʻi.

Risk of Undermining Informal Community Efforts

Ua ea kona poʻo
His head was raised

The fishing, I was watching the abuses going down out there and it was being attacked at all levels, . . . more and more outsiders. [A few years ago] it was the opening of lobster season. . . . Guys went at five or six oʻclock, evening time, setting the net six hours before the opening. And guys harvesting limu [by the] bucket! You go by the dump, . . . guys get so much they just throwing it away. Everything go to waste.
—Presley Wann, 2016

Despite formal governance rights shifting away from the local level, rural communities throughout Hawaiʻi continue caring for their fisheries in informal ways, studying changes in resource health, educating outside users and guarding against overharvest, all without state agency assistance. In Hāʻena, local fishermen would cut lay nets left unattended overnight, calling them "death traps," for any fish swimming by. They confronted community members or outsiders harvesting too much and sometimes confiscated their gear.[29] Some of the destructive fishing behaviors regulated by Hāʻena locals were illegal, while others were allowed under state law. The choice to formalize community expectations for appropriate coastal use into new, stricter state regulations was controversial within the Hā'ena community. Existing community enforcement actions were unsanctioned by the state and sometimes included illegal behavior such as deflating tires on pillaging fishermens' trucks. Some community members felt this informal approach was effective. "I know one of [the greedy fishermen]. He told me, 'You guys no can tell me [anything]' but he doesn't come down here anymore."[30] Throughout the fishery rule-making process, Hā'ena community members walked a fine line between fulfilling their kuleana to take care of their place and finding ways to be welcoming and inclusive in these efforts. Fishermen refrained from informal enforcement once rule-making began, to avoid trouble with the state and to give the process time to work. However, some worried that shifting to pursue formal legal regulations left the fishery less protected.

Monopolizing Community Time Away from 'Āina

'A'ole ho'i au e ea maluna o ko'u wahi moe
I will not go up into my bed

I wish I had a secretary, so I don't have to be here at these meetings. I could be down the beach, watching, that's my TV.
—Uncle Mac Poepoe, Mo'omomi, Moloka'i, community leader and pioneer of community-based fisheries management, 2016

Fishermen carrying kuleana into governance found themselves starting email accounts, learning to use social media, tracking legislation, traveling off island, and spending long weeks in meetings. In the early years of rule-making, the average time commitment for Hā'ena's fisheries committee members was twelve hours per month, with up to twenty-five hours per person during intensive periods. This time, all volunteer, did not include community workdays or gatherings, only meetings related to the fishing rules. Some meetings lasted eight hours on weekends when people who work all week would normally get to fish. Community members leading collaborative governance efforts in Hā'ena had less time to spend with their families, their community, and on the coast that they were working to protect.

Spending time interacting with *'āina* sustains *kuleana* to engage in governance by nurturing connections to place and providing opportunities to observe, respond to changes, and inform decisions. The practice of observing seasonal changes in resources, so honed in *kilo* and *kia'i* of the past, remains crucial to assessing abundance, adapting harvests, and shaping decisions. Climate change, particularly increases in ocean temperatures, coral bleaching, and sea level rise, make active observation of nearshore marine resources even more critical. Slow-moving policy initiatives risk becoming irrelevant by failing to address new technologies of extraction, invasive species, or evolving states of resource health.

Bureaucratic Delays

He ea hele wale aku
They were flesh, a wind that passes away

It takes some very kind of special people to have the patience. It's so methodical, just to trudge along and keep at it and keep at it. I don't know if there's a way to compare it to cultivation or to farming, or the patience it takes to fish, to help them translate this process to their world.
—Process facilitator, 2009

The Hāʻena rule-making process faced continual delays and dragged on for nearly a decade. Delays were caused by Hawaiʻi's bureaucratic Chapter 91 process, which guides creation of administrative law; by political pressure against the rules from powerful commercial fishing lobby groups; and by turnovers in state personnel. After the community submitted their formal rules package in June of 2011, the state did not respond for a year, while new appointees took over agency leadership following the election of a new governor. Frustrating delays also stemmed from uncertain and shifting legal requirements. The community received conflicting directives regarding which agencies had jurisdiction over the rule-making process, resulting in removal of multiple rules after the community had worked for five years to refine them with DLNR staff. In 2015, the rules were finally signed into law but could not be implemented without an approved management plan, which then had to be rushed. DAR staff wrote an entirely new plan and required an additional public meeting to present the new document before allowing rules to take effect.

The Hāʻena rules represented a pioneering process, leaving community members often feeling like the requirements were constantly changing. Just as they tackled one hurdle, they would suddenly be expected to jump over another. Hāʻena's eldest fisherman, Tom Hashimoto, who helped initiate the rules, described his frustration with the process this way: "My age [is climbing], I'm seventy-six years old. The delay, . . . we start working here and then something else comes up and the agenda is left on the side, it's a stand-still . . . . Then we're taking on these other problems that arise. That kills me. . . . I get frustrated."[31]

Younger community members in their teens and twenties, who were actively engaged in the early years of rule-making, dropped out. They were disillusioned with the slow pace of the process and ever-narrowing scope of the rules. Some joined pro-independence Hawaiian sovereignty organizations outside the community that oppose all efforts to work with the state. In the face of frustrating delays and restrictions, these other organizations may have appeared to be more effective governing entities.

Exhaustive Outreach and Engagement

Ea: Air, breeze

Despite opposition and challenges to collaboration with bureaucratic government agencies, Hāʻena succeeded in forwarding local-level management through an exhaustive decade-long outreach process. One government official explained the kuleana that community leaders carry to organize their community: they have to "represent the needs of a lot of different groups, and be positive, and have buy-in, and have asked the right people and include them, and, oh my goodness, there's so much that has to happen."[32] *Konohiki* once led community members by organizing collective work projects such as building *loko iʻa* (fishponds), cleaning *ʻauwai* (ditch systems) and harvesting fish through *hukilau*. Today's community leaders must organize lobbying, testimonies, petitions, and meetings, and engage in strategic planning efforts, all while facilitating community agreement. Hāʻena leaders engaged community members, government officials, and supporters throughout Hawaiʻi through face-to-face interactions, relationship-building, and vigorous debate over the content of rules.

Community Engagement through Face-to-Face Outreach and Facilitation

Wai ea: Aerated waters

In every community I've seen more processes derailed by intra-community conflict than by any other force. When I think about sustainable resource management, I think having both the value and skill for conflict management would be a really good thing.
—Process facilitator, 2009

*I know it's hard, but that's why we're here. I'm here because of you. Bringing
everybody together like this, we can come up with some solution where we can stick
together and work things out.*
—Hāʻena Hui Makaʻāinana o Makana president, 2010

Face-to-face engagement was crucial to building support for the rules, despite divisions within the Hāʻena community. After Act 241 passed, the small group of ten community members, informally referred to as the Hāʻena fisheries committee, began meeting regularly to draft the rules. This group, who added and lost members over the ten-year rule-making process, made regular efforts to gather input from various layers of the community. First, they met with the core community of Hāʻena's twenty *kupa ʻāina* ancestral Hawaiian families. As one leader explained, "The families from here first should meet before we go out and take it a step further. This is to me, something for us, for our kids. We've got to bring everybody together to see this picture."[33]

In 2008 and 2009, the fisheries committee hosted two large gatherings to discuss draft rules. These were organized like most community parties, held under easy corner tents in one area *ʻohana*'s grassy yard. There was plentiful food, including poi and fried fish, staples of Hāʻena's land and sea served at most community gatherings. Fishing committee members took fliers door-to-door to invite each family. Over sixty individuals attended these backyard party-style "meetings" to discuss the rules, representing every *kupa ʻāina* Hawaiian family of Hāʻena.

The majority of attendees at these meetings supported all of the proposed rules. Maintaining this strong core of community agreement was a time-intensive process, however, made more difficult by the ten-year duration of rule-making, state delays, changes, and compromises along the way. By 2012, there was concern that the rules had changed so much through state review that families who were originally supportive would not recognize the final version. Most Hāʻena community members did not regularly use email. In contemplating going door-to-door to speak with everyone again, one leader encapsulated the challenges Hāʻena faced, saying, "You cannot beat the small face-to-face discussions, but my concern is people don't have time. Who's going to do that? And how are we going to keep the momentum going?"[34]

One key to maintaining momentum was well-facilitated, community-organized meetings to gather input and support for the rules. In the early years of the process, DAR had no staff dedicated to CBSFA rules development or community outreach, and no staff members with relevant training. As one DAR staffer stated, "We're trusting the community's doing a good participatory process, because we don't have the capacity to run that for them."[35] The Hawai'i Community Stewardship Network's executive director facilitated rule-making meetings at no expense to the state. Four community meetings on the fishery rules were announced on the local public radio station, and held in collaboration with the neighborhood association. Hā'ena community leaders invited all users of the area to be a part of their effort, explained its history, and shared draft rules. Non-Hawaiian residents of Hā'ena and regular users such as surfers suggested changes and proposed new rules at these meetings. One woman who moved to Hā'ena to live at Taylor Camp[36] in the 1960s suggested a ban on taking live shells or diving for empty ones. This added rule prevents commercial collectors, selling shells for jewelry, from harvesting shells before they wash up on shore, leaving nothing for beachcombers.

Often at community meetings, a few outspoken individuals talk, making it difficult to gauge the opinions of the majority of silent onlookers. The facilitator used nonverbal methods to gather feedback.[37] At one meeting, forty attendees were given stickers to poll their level of support for each rule, red (stop, don't support), yellow (go slow, unsure), or green (go, I support this rule). They placed their stickers on poster paper affixed to the walls listing the rules. Over 90 percent of the votes were green. One man at the meeting, in his sixties, asked loudly, "What color do I use for go faster?"[38]

Community meetings were characterized by vigorous debate, in which people wrestled to balance protection of natural resources with fairness, practicality, and community values. Leaders took time to visit individual community members who disagreed with or opposed certain rules to find out why. Opposition from within the Hā'ena community was especially concerning because government officials repeatedly warned that a divided community would sink the process. DLNR staff argued that a strong showing of community support at Hā'ena's public hearing on the rules was critical to helping the state withstand opposition from outside forces, such as commercial fishermen. As one DAR staffer explained, "The last thing we need, when we go out to public hearing, is to have the core community saying they're not happy. . . . That just takes us nowhere. And then it makes us not want to do it, right?"[39]

Community engagement and support is critical to any local-level management effort. However, agency expectations of unified community agreement were unrealistic. Despite their years of community outreach work, Hā'ena leaders constantly faced the possibility that last-minute rule changes, outside forces, old conflicts between families, rumors, or misinformation could spark new community opposition and sink the rules package.

The Power of Relationships

Ea: Life

What I had there was certain levels of trust with key elements in that community,
so it helped. . . . I knew some key people going in and then I was willing.
—Head of DAR during passage of Act 241, 2011

Face-to-face relationship-building with government officials also helped
forward Hāʻenaʻs rule-making despite lack of state structures, capacity, or
support for devolving management to communities. State officials who had
worked with Hāʻena community leaders for a decade to restore loʻi within
the state park advocated within DLNR to extend Hāʻenaʻs community management efforts from land into coastal waters. Gaining the respect of staff
within specific agencies helped pave the way for positive relationships with
other agencies and individuals.

Despite the support of DLNR leadership, state personnel struggled with
the new CBSFA mandate that required them to work with community. In
2006, when Act 241 passed, DAR was made up mainly of fisheries biologists trained to conduct stock assessments, not write administrative rules
or engage people. As one DAR staffer put it, "We got this statute laid on us
[by the legislature], with nobody internally with the skill set to do it, or the
time. . . . To act like we just know how to do it and should figure out how to
make it happen, I feel, is crazy. In order to implement the statute effectively,
DAR needs corresponding funding and capacity." She went on to highlight
the importance of building positive relationships with agency personnel
accustomed to ire from the public. "Everybody's swamped, so who's your
champion within a division to get something moving? . . . You definitely
don't want to alienate the few government allies you have. Because it's not
an easy system."[40] Hāʻena leaders gained allies by treating staff with respect,
appreciation, and understanding, rather than blaming them for limitations
and delays beyond their control. Multiple DAR staff devoted substantial
time to forward the Hāʻena rules even when the work fell outside their job
descriptions. At times, supportive staff moved rules forward despite admin-
istrators opposed to community management.

Yet community leaders had to repeatedly build new relationships as
personnel and administrations shifted over the ten-year rule-making process. In this effort, the Hāʻena community worked with three governors,
four heads of DLNR, and five heads of the Division of Aquatic Resources,
and engaged with over twenty different county, state, and federal agencies.
Personnel transitioning between agencies also found ways to continue to

support Hāʻena from new roles within government. Trusting, positive relationships between individuals transcended agencies, job titles, and obstacles.

Hosting meetings at the Hui's taro patches within the state park, in Kēʻē, helped introduce new government personnel to community efforts, cleared up misunderstandings, and kept the rules moving toward approval after they were held up on multiple occasions. One Hāʻena leader pointed out the need to articulate the purpose of community efforts to individuals in government, universities, conservation organizations, and other partners with different worldviews: "I think a thing that is probably lacking in a lot of communities is a person who can navigate this process and translate. . . . Because obviously [communities are] doing this [local management] because of the cultural traditions associated with it, but that doesn't always come with the skill set to be able to translate between the two worlds. So I would recommend that communities don't forget about that role."[41] Even in today's world of reliance on emails, phones, and social media for communication, progress relied on face-to-face interactions and relationships to influence decision-making and build support, both within and beyond the Hāʻena community.

A visit by state and federal officials to the coast of Hāʻena with the Hui Makaʻāinana o Makana to discuss and conduct site visits for Hāʻena's fishing rules in 2012. *Left to right*: Malia Chow of NOAA, Bully Mission of DOCARE, Keliʻi Alapaʻi and Presley Wann of Hui Makaʻainana, Elia Herman of DAR, and William Aila, then Chair of DLNR. (Photo by Anuhea Taniguchi)

Valuing Community Skills and Engaging Allies

Ea: Breath

Look into what roles are needed to get a good foundation, and to see this all the way through. Then, figure out who can play each of those roles. If you figure out that there is a critical role missing, then you're going to have to figure out who you can get to help with that.
—Kawika Winter, 2009

In order to engage in governance, Hā'ena community leaders had to work within a system that did not formally recognize their role, expertise, or commitment of voluntary time. Their efforts required multiple, varied skill sets quite different than those of a skilled farmer or fisher. Leaders engaged community members to contribute in diverse ways, from cooking meals for workdays and meetings to monitoring the health of marine resources in the ocean, keeping sign-in sheets to calculate volunteer hours, and tracking state and county legislation. Hā'ena leaders also recruited outside partners to fill skill sets that did not exist within the community. Individuals with diverse skills were welcomed to contribute. Kawika Winter encapsulated Hā'ena's strength in diversity this way: "I think having diversity on different levels helps to push us forward. You have *kūpuna*, *mākua*, warriors, fishermen, farmers, researchers, college graduates, all coming together and coalescing like the *'aha* [braided cord], stronger than a single cord would be. I think that's one thing Hā'ena has to its advantage."[42]

After all of these extensive efforts to engage diverse groups of users and obtain input, support for Hā'ena's rules was substantial both within and beyond the community. Over two hundred people attended the public hearing in Hanalei, held October 2, 2014. Every one of the area's Hawaiian families stood up in groups of four to ten individuals, two and three generations strong, to declare their support. Uncle Presley recalled, "I can't think of a single family from Wainiha to Hā'ena who wasn't there represented. It was just beautiful."[43] Non-Hawaiian Hā'ena and Halele'a residents, including the president of the Hā'ena-Hanalei Community Association, testified along with regular users including surfers and kite boarders, students, researchers, and *kia'i* from communities all across Hawai'i, in a powerful and diverse showing of support. One non-Hawaiian resident of forty years shared, "I'm a *malihini* [visitor] and I will probably always be one. I came here about forty years ago to Wainiha and . . . I'm astounded at the *mālama 'āina* [caring for land] and the number of people that came out tonight to support Hā'ena and the plan they've created." He went on to describe growing up in Washington State, between the ocean and bay, when clams were in abundance and you could dig them up and eat them, the water was so pure: "By the time I moved away, the signs were there that said don't eat the clams; they're poisonous. . . . You folks are in a position to stop that process."[44]

Another longtime non-Hawaiian community member explained, "I don't have the expertise on details [of the rules], but I trust the knowledge, the reputations, the credibility, and the love for the land and the waters of local experts and *kūpuna* who have worked on the development of the proposed rules. I feel very comfortable supporting their recommendations, and I urge you to move forward with implementation as soon as possible."[45]

Disappointments and Barriers

Ea: Rule (to govern)

If the state could just see how hard people are trying to work with them, trying so hard to work with a state system that is so broken.
—Debbie Gowensmith, Hāʻena rule-making process facilitator, 2009

Despite the community's exhaustive engagement process and success in attracting diverse partners and supporters in government, Hāʻena still encountered disappointments in the process and outcomes of rule-making. The final Hāʻena rules passed into law were much narrower in scope than the community's original draft proposals, and likely less environmentally protective.[46] Some key barriers were fragmented state resource management agencies with narrow jurisdictions and the inability to adapt rules to real conditions.

The Hāʻena community aimed to increase fishery health by addressing holistic mountain to sea threats, believing that the health of the ocean begins on land. However, DLNR assigns specific jurisdiction over fresh water, forests, coastal lands, recreation, and aquatic resources to four different agencies. Though the original CBSFA legislation appointed DLNR to work with communities to develop rules, the department delegated the process to its subsidiary agency, the Division of Aquatic Resources (DAR). Working through the DAR precluded rules related to terrestrial impacts such as coastal cesspools or decrease in fresh water flowing to the ocean. A number of rules focused on reducing user conflicts, and the impacts of high-volume recreational use were also left out, including limits on boat size, use of environmentally damaging sunscreens, and separate zones for fishing, snorkeling, and other recreational activities. DLNR argued that these rules fell under a separate DLNR subsidiary agency, DOBOR, The Division of Boating and Recreation.[47]

As a result of DLNR's narrow jurisdictional focus, the final community rules, now law, address mainly fishing, despite many other identified threats to coastal resource health. Consequentially, newer opposition to the rules has emerged from within Hāʻena and Kauaʻi communities, concerned that the law ultimately regulates only the twenty fisher men and women who regularly use the area, while ignoring the impact and activities of nearly one million tourists per year. Whereas *konohiki-* and *ahupuaʻa*-level governance integrated a range of impacts within a single decision-making unit from mountains to sea, the organization and scale of state agencies stymies holistic, ecosystem-based management in contemporary times.[48]

Another conflict was state agencies' lack of flexibility to adapt rules or craft them to reflect change and complexity in ecosystems. The states' administrative rule-making process, Chapter 91, requires multiple rounds of review by the public, the attorney general, the Small Business Administration, the State Board of Land and Natural Resources, and the lieutenant governor and governor before changes can be implemented.[49] Conversely, customary local-level Hawaiian fisheries management was adaptive. *Konohiki* and *kilo* opened and closed species and fishing areas in response to real-time observations of resource health.[50] Fisheries committee members proposed flexibility to adjust rules: for example, opening closures on species once they were observed to recover. But DAR staff held that any proposed changes would have to repeat all of the steps of Chapter 91 review. In response, community members opted to close less species than they had originally wanted to rest, not knowing whether it would ever be legal to fish them again.[51] Lack of flexibility for review and adjustment raises the stakes for proposed rules, discouraging experimentation, learning, and responsiveness to actual conditions.

Challenges of Balancing Public Access with Community Control

Kaha ea: To deprive of rights of livelihood

If you want to set up a set of sustainable practices that apply to you and everybody else in this place, then that will fly. But you don't get special treatment.
—Former head of DAR, 2011

I think we need to admit there are levels of stakeholders within this. And I don't know how we do that legally, but there are, right?
—DAR staffer, 2011

In the Hawaiian Kingdom, fishing regulation was local, and access to fisheries was limited to local residents. In contrast, Hawaiʻi State law affords local community members the same protections, rights, and access as any member of the general public. DAR staff and the attorney general's office vetoed a number of proposed community rules on the grounds they might violate Hawaiʻi State constitutional protections on public access to the coast, from the shoreline to the highest annual wash of the waves.[52] One proposed rule would have required anyone fishing in Hāʻena to obtain a permit by passing a course focused on acceptable fishing practices and the significance of the area, just as hunters must pass a class to hunt on state lands: "Anyone who wanted to fish in Hāʻena would have to get a special permit. It wouldn't cost them anything, but it would require that they be educated about the rules, . . . read and acknowledge the rules. . . . That was found unconstitutional, so we took it out."[53]

Another proposed rule would have required boats fishing in the area to launch from Hāʻena, giving community members an opportunity to talk with boaters, explain area rules, and check their catch. This rule was submitted in the formal rules package, then removed after attorney general review ruled it an unconstitutional limit on public access.[54]

To meet state-level constitutional standards of equal protection, rules could not elevate the rights of certain user groups over others. One proposed community rule formalized the longtime local practice of limiting harvest of *limu* on the most accessible area reef to elders, protecting an area where elders can safely gather food for themselves.[55] DAR staff asserted that this rule was discriminatory, because it limited access based on identity, creating distinct protections for a particular age group. This interpretation undermined cultural standards of esteem honoring elders as a distinct group for their wisdom, teaching, and years of community contributions.

Rules also could not apply differently to tourists and residents, including residents of Hāʻena. In order for proposed regulations to withstand legal scrutiny and become state law, any activity allowed for residents had to be allowed for everyone, including tourists, though there were more than one hundred times as many tourists using the area.[56] Though certain activities posed no threat to resource health when carried out by only a handful of community members, Hāʻena families chose to give up practices they had long enjoyed together, in order to regulate such uses by the general public: "We've all got to give up something. [Our family] enjoys snorkeling right offshore for shells when the crowds die off. But if that's something we have to give up so the commercial guys can't come through, then that's that."[57]

Here *kuleana* encompasses making sacrifices so the resources can recover, leading by example.[58] As Uncle Jeff Chandler stated, "If we don't set limits for ourselves, we can't limit anyone else."[59] Hāʻena fishermen offered to give up fishing a lagoon that offered the most protected and easiest access because it was a key nursery habitat. They also wanted to close this area to other activities they felt disrupted feeding and spawning in the nursery, including snorkeling, swimming, kite boarding, and surfing. In a 2009 meeting with recreational users, Uncle Jeff Chandler implored: "We all impact the fishery. I impact because I eat them. [Fishing] is my life, how I was born and raised, what I want to pass on to my son. This is our best fishing ground we're giving up. That is what I'm going to sacrifice, and I ask you to do the same thing. . . . What are we all going to give up for the benefit of the future?"

However, not all users were equally willing to sacrifice. Requiring the community to get all stakeholders to agree on each rule submitted for government approval weakened protection of resources. Surfers argued that they needed to paddle through one section of the lagoon to reach a popular outer-reef break. Kite surfers opposed closing a channel where turtles and certain fish feed because reaching the next safe landing point would add forty-five minutes of upwind sail time before they could reach shore. Removing these areas to achieve consensus with recreational users whittled the final protected area down to a size that may be too small to produce expected ecological benefits such as spillover, in which fish spread to enhance abundance on other parts of the coast.[60]

One longtime participant in rule-making expressed her frustration that local fishermens' recommendations for the larger protected area were not adopted: "It's a *subsistence* fishing area. Can't we rank who gets a say? [The state is] creating areas with different purposes [recreation, commercial, subsistence]. If everyone has an equal say, you might as well just create one kind of area."[61]

In the Hāʻena case, rules and environmental protections would have been stronger if the process had a means to differentiate rights for subsistence fishermen, community members, and elders from those of the general public, in keeping with traditional practice and the goals of the CBSFA. Instead, the process protected equal access for harvest and use, along with equal input into rule-making.

Statewide Public Decision-Making

Hōʻea: To arrive

Why should someone on [the island of] Maui get to have a say about what happens in Hāʻena?
—Debbie Gowensmith, 2009

When I put out a public meeting notice, it is for the public in general. . . . I am not overly concerned with who and what is community.
—Former head of DAR, 2011

The Hāʻena community had to undergo three separate public hearings to gather testimony in order to secure their CBSFA designation and rules. Commercial fishermen and tour company operators on Oʻahu and Maui submitted testimony against Hāʻenaʼs initial CBSFA designation, though these individuals had never set foot in Hāʻena. Both groups opposed passage of Hāʻenaʼs rules on the grounds that similar place-based restrictions on commercial coastal use could spread throughout the state.

Government decision-making can be manipulated by powerful interest groups despite the stateʼs emphasis on due process and equal consideration of all stakeholder input.[62] Lobbyists filed two contested case lawsuits to stop the Board of Land and Natural Resources (BLNR) from voting on the rules. One claimed that a fisherman in opposition was given less than the three-minute time allotted for any public testimony. Video footage of public hearings disproved this accusation, and BLNR members unanimously voted to throw out both lawsuits and approve the rules. The same lobbyists then worked with commercial fishermen on the west side of Kauaʻi, twenty miles from Hāʻena, to convince a powerful Kauaʻi legislator to keep the governor from signing the rules into law for another nine months. While those who benefit commercially have powerful avenues of influence, there is no legal means of weighting input from those who are most connected to and knowledgeable about resources, who actively contribute to their caretaking.

Achievements and Lessons

Ke ea ʻana o ka ʻai, ka iʻa: The obtaining of poi and fish, the ability to feed oneʼs family, self determination[63]

The many challenges in the rule-making process led to disappointments, including narrow rules that lost many of their creative and customary aspects and ended up looking a lot like common state-level fishing regulations (appendix B). However, in working to pass rules, the Hāʻena community achieved far more, capturing the imagination and support of diverse groups across Hawaiʻi through their tireless work and sacrifice to care for their home. In this final section of the chapter, we focus on some of the key achievements and lessons from this process.

Building Local-Level Capacity for Independent Governance

Ea: Autonomous local governance

I think sovereignty will be easier than this. Anything will be.
—Debbie Gowensmith, facilitator, 2009

How do we get the things we see as constraining us, and find a way to use them to our advantage?
—Chipper Wichman, 2011

Hāʻena community efforts show how it is possible to work within existing imposed state structures, while simultaneously rebuilding systems of local-level governance. One former DAR administrator articulated the need for this work:

> The best [community members] can really do [right now] is to call up [the DAR office and] say we have a problem out here. Could you please send the biologist? . . . You might or might not get an acceptable response on the given day, and even if you did . . . someone had to drive all the way around the island to get there. . . . I think what you need is some way . . . to fill out local institutions . . . to promote some locally empowered form of resource management. Sort of quasi-build the *konohiki* system.[64]

Like Hāʻena, communities across Hawaiʻi are engaged not just in environmental protection, but also in strengthening their capacity to govern, independent of the state. Over the past twenty years, Hāʻena community leaders have slowly built their community group, the Hui Makaʻāinana o Makana, to serve as a vehicle for local governance. The Hui began holding informal monthly workdays at the *loʻi* in 1998. Starting in 2011, instead of meeting separately, the fisheries committee began using these regular Hui workdays as a chance to give updates, gather input, and make community decisions on the fisheries rules. By 2012, the Hui president and vice president began to serve as spokespeople for the rules and, through them, other area issues. As the Hui and its network of area Hawaiian *ʻohana* grew more active and organized, the fisheries committee was less needed and, by 2015, largely stopped meeting. Instead of having to leave Hāʻena to represent their community at multiple different meetings on specific topics with agencies that do not interact, Hui leaders encourage agency staff and other partners to come to workdays. This way, related issues are addressed all at once, at gatherings Hāʻena families already regularly attend together. At work sessions, visitors interact with community members as their hosts, rather than as "interested parties" encountered for a few hours in a hearing room. An average Saturday Hui workday includes twenty community members, along with representatives of multiple state agencies, students assisting with community-led research projects, and often a visiting school or group of cultural practitioners coming to learn. Workdays engage all participants in helping to care for Hāʻena together. People talk while pulling weeds from the mud of the taro patch, share lunch, get to know one another in a comfortable setting, and make decisions about Hāʻena's resources on the land.

By building community organizations and strengthening relationships among community members, outside partners, and the land, Hāʻena leaders are setting a foundation for a time when the community can again govern itself. One NGO leader who has supported Hāʻena's work described the community's approach by saying, "Start creating change in the system you have; don't just fight for a new system."[65] Though the Hāʻena loʻi are still "owned" by the state, Hui members whose families have farmed these same taro patches for generations are the hosts, decision-makers, and cultivators whose collective work is visible on the landscape. To enter the loʻi is to enter a space of Hawaiian community autonomy, a space in which to perpetuate practices of the past and envision the future.[66] As one supporter said in 2009 of their gradual efforts, "This isn't sovereignty, it is just a baby step, but sometimes it takes those to get what you want in the end."[67]

The Power of Networks

Ea: To Rise

I agreed to do outreach for my community—I had no idea what I would be getting into. Now I'm here with all these other communities.
—JL, Hana, Maui, community leader, April 2016

Hāʻena addressed powerful organized opposition to the rules from statewide lobbies by mobilizing their own network of support. Working with KUA and the E Alu Pū network of more than thirty-two different community groups engaged in similar efforts, Hāʻena built allies and political strength across Hawaiʻi. Community leaders worked for years to secure a hearing on Kauaʻi, in Hanalei, at the closest public meeting space, twenty minutes from Hāʻena, where hundreds testified. However, actual decision-making took place two weeks later at a hearing of the Board of Land and Natural Resources, held in the state capital of Honolulu, making it expensive and prohibitive for community members and actual users of Hāʻena to attend. Only ten Hāʻena community members could make the trip. However, sixty-three people—from the University of Hawaiʻi, nonprofit conservation organizations, and communities connected to Hāʻena through KUA and E Alu Pū—showed up to testify in support of the rules. Only two testimonies were in opposition. Working within a lager movement of communities working for local coastal management, Hāʻena leaders never had to face decisions alone. They had partners with whom to strategize, problem-solve, and craft a common message to benefit all.

Lawaiʻa Pono: Fishing *Pono*, Responsibilities before Rights

Ea: Spirit

> *It's not about pointing fingers. It's something that we learned from way back when we were small: mālama what you get, take care what you get, take what you need and that's it, think about tomorrow, think about the future. Simple. So remember every one of you in this room get something to do with this.*
> —Keliʻi Alapaʻi, 2016

Despite stringent legal screening to ensure that Hāʻenaʻs rules protected public access, opponents of the rules spread the misconception that no one from outside Hāʻena would be able to fish the Hāʻena coast. In newspaper editorials, fishing magazines, blogs, and testimony, commercial fishermen argued that Hāʻenaʻs rules violated their "rights to fish" anywhere across the state. A handful of Native Hawaiians from outside Hāʻena argued that the rules violated state constitutional protections on traditional and customary gathering rights. One Hawaiian leader suggested he would support the rules as long as they exempted all Native Hawaiians.[68]

The Hāʻena community could have argued against these inaccurate claims to gathering rights, citing Hawaiian Kingdom law, *konohiki* and *ʻohana* fishing practices, and how certain fishing spots were customarily reserved for and cultivated by area families. Instead, they chose to focus on the positive, shifting from exclusive *rights* to inclusive *responsibilities*. In newspaper interviews and face-to-face meetings with fishermen around Kauaʻi, the president and vice president of the Hui repeatedly stressed that everyone would continue to be welcome in Hāʻena. As Uncle Keliʻi Alapaʻi, vice president of Hui Makaʻāinana o Makana, said in publicity on the rules, everyone was welcome to fish in Hāʻena, "We are simply asking that when you fish in our community, you respect our traditions and the way we fish."

Together, in support of Hāʻena, the E Alu Pū network developed a positive Facebook campaign, T-shirts, and messaging around the phrase "*lawaiʻa pono*," to fish in a fair, correct, and balanced way. *Lawaiʻa pono* encapsulates *kuleana* described earlier in this book, including taking only what you need, observing reproductive cycles, taking care of fishing spots before you gather from them, respecting the people of a place and its resources. In the Hāʻena public hearings, Hawaiian fishing families and their supporters reclaimed the meaning of *lawaiʻa* (fisherman), and the idea within that word of always ensuring *lawa* (enough) *iʻa* (fish).[69] They offered a model of fishing, and of being Native Hawaiian, based on responsibilities rather than rights. *Real lawaiʻa*, Hawaiian or not, do not need destructive gear and do not overfish or raid other people's fishing grounds. Instead, they exercise their *kuleana* to care for, stand up for, and protect the reefs that feed them, inspiring others to do the same. Since, Hāʻena's rules became law, the community of Kaʻūpūlehu on Hawaiʻi Island passed their own community rules

Harry and Makaʻala Kaʻaumoana of Hui Makaʻāinana o Makana in Hāʻena. (Photo by Anuhea Taniguchi)

in 2015, a straightforward ten-year ban on all fishing along their coast. In 2016, Moʻomomi, the community that pioneered the first CBSFA effort on Molokaʻi, entered the public process for approval of their own community-based rules, using the same motto, *lawaiʻa pono*.

Enduring Commitment

Aea: To come up from underwater

The amount of human will and passion that has been put into what we have is immense. It's immense, the cost. I'm not talking about the grants; the cost to people is huge, and so you want the outcome to be worth it. This is why I don't sleep at night.
—Debbie Gowensmith, 2009

The level of engagement of government agencies and conservation NGOs in Hāʻena shifts with policy initiatives, personnel, and funding cycles. However, community commitment to place is enduring. Throughout this thirty-year process, founding members of the Hui and fishing committee continued to show up at Hui meetings, meet with elected officials, and work on the fisheries management effort. These leaders persisted despite transitions in their own lives, whether or not they continued to hold formal elected leadership roles. In the face of frustrations with the slow pace of working with state government, and the sense of continually having to start over, community members never gave up.

When asked how they kept going all these years, Hāʻena community members answered with *kuleana*—*kuleana* to their place, to other communities engaged in similar efforts, and to future generations. One key source of endurance for community leaders was reconnection to the land and ocean, which motivated their engagement in the first place. Work days on the land, and time to fish or bring their grandchildren to the places they care about, all helped keep people going. Collaboration with groups beyond the community, particularly through KUA, provided perspective, inspiration, and a sustaining force. Hāʻena leaders were keenly aware that passage of their rules would set precedent for other Hawaiʻi communities engaged in similar efforts, including those whose leaders had become friends through the E Alu Pū network. Even at the lowest moments, giving up was simply not an option because of their commitment and *kuleana* to other communities. A final key motivator was working for future generations, such as their own grandchildren, to have a connection and sense of *kuleana* to place: "We all have to come together and put our minds together to take care of what was given to us by the generations before us, so we can keep what we have for new generations that are going to come."[70]

Kiaʻi: Local Guardians

Ua mau ke ea o ka ʻāina i ka pono
The life of the land is perpetuated in righteousness

It's the coming together and saying, "Hey we want our culture to live." It's obviously not how it was two hundred years ago, it's not how it was one hundred years ago, but we're going to figure out how to operate within this system to make sure that our culture survives today. We can still eat from the land and the ocean and mālama ʻāina so that our families are healthy. And so that's the role, it is the embodiment of the community's voice that is attempting to be resilient in the face of change.
—Kawika Winter, 2011

Community leaders in Hāʻena had to cultivate new awareness and skills to carry kuleana into governance, responding to constant changes both in the land and ocean of their home and in the social and political landscape. The word *kiaʻi* is active, connoting vigilance, ongoing watchfulness, and *maka ʻala* (awake eyes).[71] A *kiaʻi loko*, the caretaker of a fishpond, watches for changes in the fishpond. He or she observes the flow of water in and out of the *mākāhā* (entrance sluice gate) and the fish gathering on the rising and ebbing tide. *Kiaʻi loko* lived at fishponds, often in hale located right next to the *mākāhā*, so that they could catch poachers coming to steal fish at night, respond to changes in weather, and open and close the gates. *Kiaʻi loko* (within) also had to be aware of changes within themselves and their community. *Kiaʻi* of today extend their vigilance to tracking bills at the legislature and monitoring commercial fishing blogs, Facebook posts, and actions at the neighborhood board.

Community *kiaʻi* work to overcome barriers in state and county government structures that stymie collaborative efforts. One key barrier is the inability to make decisions about resources holistically, rather than in fragmented agency jurisdictions that separate fresh water, forests, and fisheries. Another need is the ability to adapt and change rules in response to real-time changes and complexity in resources and social conditions. Finally, new institutions must find legal means to recognize *kuleana*, allocating rights based on fulfillment of responsibilities. *Kuleana*-based governance would weigh sacrifice and contributions to a place in resolving conflicting interests, uses, and policy input.[72] Examples of *kuleana*-based decision-making include

- Reserving limited catch allocations for subsistence fishers who share their catch.

- Closing public access to a state park one day a week, limiting entry to community members helping with restoration projects.
- Allocating commercial permits for use of public lands based on proposed stewardship efforts such as hauling out trash, removing invasive species, or providing education to tourists.
- Weighting public response to a development proposal or management plan based on criteria such as the number of meetings a respondent has attended, how regularly they use the affected area, and the degree to which their use provides community versus private benefits.

Collective rights based on responsibilities are crucial to nurturing community caretaking and governance. Examination of the Hā'ena case argues for devolution of authority for governance to those who care most about a place. Local-level decision-making would likely be more stable and effective than distant, fragmented agencies overseeing the entire state. The Hā'ena rule-making process promised shared decision-making authority with state government. Instead, after investing years of their time, community members had to drop multiple proposed rules. They still lack formal authority to enforce those that passed. There is no means to recognize community input or effort by differentiating community members from any other members of the public. And the boundaries around who is included in "community" are unclear, requiring leaders to consult with ever-broader groups. Yet, despite all of these disappointments and shortcomings, Hā'ena persevered in pioneering community-based coastal management rules rooted in Hawaiian ancestral values and practices. More important than any rules, Hā'ena captivated imaginations and diverse support by showing all of Hawai'i what it means to be truly committed to protecting and carrying *kuleana* for one's home. In doing so, community leaders helped to catalyze a Hawai'i-wide movement rooted in *kuleana* to protect and care for beloved, sustaining places.

Community leaders of Halele'a, like Hawaiian leaders across the state, are engaged not just in resource protection, but also in institution building, as they shepherd local-level organizations and enhance capacity for restored local governance. They are engaging diverse community members and allies in building *ahupua'a*-based networks of *kia'i* across Hawai'i. Though they must work within existing power structures and imposed state and federal governments of the United States, they are simultaneously cultivating collective, *kuleana*-based independence.

Mo'olelo: Hālawai (Hā'ena Public Hearing)

Hanalei Elementary School, October 3, 2014, 6:30 P.M.–9:00 P.M.

The parking lot at Hanalei School is full. I park across the street, past the church. People flow across the lot, talk in clumps on the grass, and spill down the cafeteria steps as they wait for the doors to open and the testimony sign-up sheets to be set out. The crowd is wearing off-white T-shirts that say "Lawai'a Pono" in purple, looking like an off-white waterfall cascading down the stairs scattering droplets. Uncle Keli'i Alapa'i is talking with an uncle from Moloka'i, the rain-washed face of Mount Namolokama spilling real waterfalls into Wai'oli Valley just above their heads. People point me toward the covered walkway for parent dropoff, where a mural on the wall tries to capture the same mountain view. Uncle Presley's wife, Aunty Colleen, is giving out T-shirts. "Medium?" she asks, marking my name on her clipboard, "An extra-large for your husband, and what size for your mom and sister?" I walk away with a pile.

Those assembled for the hearing gather at the foot of the steps, hold hands to form a long oval aligned mountain to sea. Opening protocol must take place before official hearing start time, and outside the building: separation

People arriving for the public hearing on Hā'ena's community-based fishing rules and waiting for the cafeteria doors to open at Hanalei Elementary School on October 3, 2014. (Photo by Kimberly Moa, KUA)

Opening community-led protocol outside before the public hearing began. (Photo by Kimberly Moa, KUA)

of church and state. One of Uncle Jeff's nephews, Kamealoha Forrest, a practitioner of hula, offers an *oli* that honors Hāʻena, calling on the ancestors of that place to join us just down the highway here in Hanalei. His great-uncle Samson Mahuiki offers the prayer. The circle of two hundred is too big for most of us to make out the words, but we feel the slow rumble of his voice. When we enter together, we fill the cafeteria. Every folding bench is packed. People stand against the walls, and three to four deep on the surrounding *lānai* (porch), leaning in through the open doors. Our children play outside on the grass within view. I see everyone I know and many faces I don't. Those who do not have official T-shirts have come wearing off-white, all but five people in the crowd now grown to maybe three hundred. When the lead paid lobbyist for commercial fishermen opposing this bill arrives, he is given hugs and a T-shirt. An hour into the testimonies, he leaves without testifying.

The acting administrator of DAR welcomes us in a crisp official voice. He tells the crowd that DAR will not be answering questions on the rules tonight. This is a public hearing for the sole purpose of taking public testimony. Every speaker will have three minutes. A buzzer will sound at two and a half, and the administrator will cut us off when the second buzzer sounds at three. We must be respectful of everyone signed up to speak. We are to testify toward him, sitting alone at a long table in front of the room. We are not to face or interact with the crowd.[1] To begin, the administrator reads the rules verbatim in their official legal language: "Any person who violates any provision of this chapter or the terms and conditions of any permit issued applicable to this chapter shall be subject to administrative fines of not less than $100 and not more than $1000 for a first violation."[2] Hearing the words said in his voice feel so sterile, menacing, and full of prohibitions that I am afraid many in the crowd may change their minds on the spot.

Indeed, the first speaker, an enforcement officer for the Division of Conservation and Resources Enforcement (DOCARE), testifying as a private citizen, argues against the rules. Buddy Wilson believes he should be allowed to bring his commercial fishing vessel into the closed lagoon area to fish. He points out that he fishes not only native but invasive species, which, if left unchecked, will take over not only the protected area, but the whole coast. He ignores both buzzers and keeps talking. The administrator, whose credibility and ability to manage the entire meeting could easily be lost in this first five minutes, tells Buddy time is up. After a few reprimands, the administrator finally succeeds in getting him to step back. During their exchange, someone catcalls and yells angrily from outside the building. The audience begins to rumble. Other hearings have broken up this way, with commercial fishing "plants" yelling insults at speakers and state officials, confusing and angering the audience as a tactic to force hearings to be shut down. The Hāʻena community has prepared for this. Young, strong, level-headed community members walk the perimeter of the crowd in pairs, talking calmly with everyone, identifying the catcallers and letting them know they must stop or leave.

The next speaker is the oldest fisherman of Hāʻena, Uncle Tommy Hashimoto. He reads his official-sounding testimony tersely, looking uncomfortable. But at the end he says, in his own words, "This makes me feel that all the hard work and effort of so many people was worth it, to be here for this occasion." Three of his daughters, two grandsons and his wife all

Aunty Lahela Chandler Correa testifies before the acting head of DAR, with her children, Keaki, Alakaʻi, Pohai, and Lahela. (Photo by Kimberly Moa)

stand up with him. One at a time, each family of Hāʻena stands up behind one family member designated to speak. They stand with their *kūpuna*, hold their *keiki*, three and four generations strong, in groups from two or three people to thirteen. They address the microphone sideways rather than facing their backs to the audience so that they can speak to the DAR administrator and their community together. Their presence is powerful beyond words: the visual of the families of a place standing in turn, the chance to see how people are related in ways not previously understood, their sense of strong and unbroken *kuleana* to this land. They shift the tone of the hearing.

After two hours and sixty-five testimonies, the meeting ends. Sixty-two testimonies were in favor, one was undecided, and two were opposed. Only a handful refer even tangentially to the actual content of the rules. Most support the families of Hāʻena and their efforts to care for their home. There are testimonies from every island, seven UH graduate students in marine policy and ecology, one kayaker, the president of Hāʻena-Hanalei Community Association, a kite surfer, and multiple Hāʻena-area residents. By the end of the night, the cafeteria is no longer full. People have filtered out, needing to put children to bed and prepare for the next day. The outcome is anticlimactic. The administrator ends the meeting, saying,

> Thank you all for making the time to be here tonight and for your testimonies, your overwhelming support of these rules and the Hāʻena community, and your respect for this process. I have run many public hearings on fishery issues, and I can tell you that none were ever this collegial or this beautiful. Your testimonies will be submitted to the members of the board, who will meet in two weeks to consider whether to pass the rules.

The remaining Hāʻena families and community members of the E Alu Pū network, who have flown from across Hawaiʻi to testify tonight, pile into pickup trucks and vans. The night air is cool as they drive the winding two-lane road back to their tents at the *loʻi* in Hāʻena where they began the day weeding. Some Hāʻena family members have left the hearing early to cook, so dinner awaits. It is 10:00 p.m. Members of the group pull out guitars and ukulele, singing together under the stars until late at night, celebrating a good day. At midnight, a rainbow circles the full moon for all to see.

NOTES

1   Meetings related to fishing in Hawaiʻi are often highly contentious. Unruly behavior, and yelling, are common, and some meetings have been shut down. State agency staff were concerned about the possibility of similar outcomes in this hearing.

2   Hawaii Administrative Rules Title 13, DLNR, Subtitle 4 Fisheries, Part 2 Marine Fisheries Management Areas (http://dlnr.hawaii.gov/dar/files/2015/08/ch60.8.pdf).

CHAPTER 7

Kaiāulu

Provisioning Community

Some people consider kuleana to be the responsibility or tasks given to you, something inherent or passed on. . . . Kuleana, I think, is something that steers your responsibility, to make sure what you are doing is leading your wa'a [canoe] in the right direction. If you get off track, if you stray from the legacy of people that took the time to teach you, then you need to make those adjustments to get back.
—Billy Kinney, age thirty-one, Hanalei, Kaua'i, 2016

In this book I have shared stories of *kuleana*-based communities in Halele'a and Ko'olau, Kaua'i, in hopes that area families' experiences might offer learning for other communities in Hawaii and around the world struggling to maintain their roles as caretakers. How can communities perpetuate *kuleana* to the places they love amid significant encroaching forces? What responsibilities come with living in a place? In an ever-changing world, how can communities grow within the bounds of places that sustain them, in ways that sustain these places in return?

I end this book with eight provisions for cultivating community caretaking and governance gleaned from these stories of Hawaiian fishing communities of Halele'a and Ko'olau, Kaua'i.[1] Every community is as unique as the place that shapes it. These provisions are neither recipe nor checklist, but lessons emerging from this work that might be of use in other settings. To show the continuity and contemporary nature of *kuleana*, I bookend each provision with quotes from young Halele'a community leaders in their twenties and thirties. Some of these young people are grandchildren of elders whose voices are foundational to earlier parts of this book. Many now have children of their own.

These provisions build on and reinforce not rules but relationships that bind people to place and to one another across generations. Rooted, woven relationships strengthen community ability to care for beloved places. Community commitment can endure even as landowners, administrations, agency personnel, conservation interests, and funding priorities shift. The lessons offered here draw on the multiple forms of *kuleana* to land and re-sources described in this book: respecting resources as both deities and kin, redefining resource use and land ownership based on caretaking and respon-sibility, cultivating collective ability, sharing to create abundance, protecting land to maintain spaces of presence and connection, restoring local gover-nance based on ancestral values, and preparing future generations to carry on.

1   *Āina:* Let Land Lead and Sustain

> *Don't forget to ask the place.*
> —Kamealoha Forrest, age twenty-nine, Hāʻena, Kauaʻi, 2016

Land and resources are living entities imbued with spirit and power. In this book, community members have shared their connections and family ties with particular resources. Deities embodied in *ʻāina* actively guide fishing and daily life. Stories emphasize the importance of building relationships and reciprocity with other living beings. Including ceremony when work-ing with *ʻāina*, creates focused time to honor a place and all those who have used and cared for it in the past: to reflect, observe, learn, and ask for guid-ance, whether through chant, prayer, or simply stillness.[2] Hawaiian scholar and master hula teacher Puanani Kanakaʻole Kanahele describes ceremony as "procedure that allows us to reach out into the unseen, but most impor-tantly, allows them to reach back to you."[3] Potowatomi botanist Robin Wall Kimmerer writes that ceremony marries the mundane to the sacred and "is a vehicle for belonging—to a family, to a people and to the land."[4] Ceremony acknowledges life-giving elements of our Earth and invites these entities to guide and strengthen efforts to protect them.[5]

It is also crucial to make decisions about a place while physically pres-ent in that place. Wherever possible, meetings should be held on the land, so that participants can see and feel real conditions.[6] Management plans are often made in distant conference rooms aided by remotely sensed data and models, scheduled around grant cycles rather than lunar calendars or seasons. In contrast, regular workdays and field trips help diverse partici-pants who collaboratively manage resources to become reacquainted with places together.[7] Working slowly—for example, clearing land with hand tools rather than bulldozers—allows decision-makers to pay close attention to the effects of their actions on the land, to learn, adjust, and let the place lead.[8] Ecosystems today face unprecedented levels of stress and uncertainty. Regular hands-on work with *ʻāina* strengthens awareness of change and, in turn, enhances adaptive capacity.[9] This work grounds decision-making in ongoing interaction with the land itself.

Land also nourishes efforts to protect it, which can be daunting. Working with government agencies to improve laws or behaviors affecting 'āina requires enormous dedication of time, ongoing outreach, and compromise to achieve small and easily reversed gains.[10] Vital formal policy and enforcement work need to be balanced by informal on-the-ground actions such as education and restoration.[11] Feeling resources thrive provides sustenance to keep efforts going. Young people cannot pick up the torch if they see only burnout.

> When I go to a meeting, I feel like I didn't do anything productive, I'd rather do a community workday and get something done, do something pono [right, just, balanced] for that place.
> —Olena Molina, age thirty, Kīlauea, Kaua'i, 2016

2   *Pono*: Cultivate Reciprocity

> If you really love a place and you are really tied to it, a locked gate will never keep you from it. . . . That place will always have that connection to you and no matter what happens, you will fight for it. That's what it means to be tied to it. . . . I'll understand other people's fight and can kako'o [support] them through . . . being connected to my place. [Do] not just run around and try and protect all of Hawai'i. Protect your place. If everybody protected their 'āina, we would be in good shape.
> —Kamealoha Forrest, Hā'ena, Kaua'i, 2016

Hawaiian relationships with place, like those of other Indigenous communities, are based on work and mutual caretaking.[12] Rights depend on responsibilities to sustain ecosystem health, the foundation of both identity and culture. These responsibilities must be taught and practiced alongside efforts to protect unique Indigenous rights rooted in the interdependence of culture and place. There is no right to harvest everywhere, only in those places we invest time caring for. Families gather in places they have cultivated over time, through planting, weeding, protecting spawning seasons, and creating habitat.[13] Underlying all these activities is the principle of reciprocity, with place and with other people. This principle is not an even accounting, in which returns equal withdrawals; it is contributing more than you take as a way of life.

For community governance to be effective, rights to use and govern natural resources must be founded on fulfillment of responsibilities. Authority should come with work that feeds place and community. Conflicting interests and claims can be resolved based on how competing groups contribute to taking care of a contested place. Where groups exercise specific responsibilities to enhance resource health, policies should recognize their distinct use and management rights.[14] To effectively care for *ʻāina*, communities require recognition of collective rights, collective harvest areas, and collective decision-making bodies that are not open to the uncommitted general public. Such privileges should be based not solely on identity, blood, or length of residence, but on understanding and practice of *kuleana*.

> *The things we are able to enjoy, the places, is because somebody worked really really hard to preserve that place.*
> —Olena Molina, age thirty, Kīlauea, Kauaʻi, 2016

Community members in our *kiaʻi ʻāina* class, one of the first courses for college credit taught in Haleleʻa. (The nearest community college is an hour and a half away). In this fall 2014 session, long time community land use experts came to teach. All are holding a historic map of the Wainiha Hui lands. *Top row, left to right*: Wayne Palala Harada, Debbie Grady, Tamra Martin, Aunty Barbara Robeson, Uncle Keith Nitta, Johanna Gomes, Kaʻimi Hermosura, Aunty Jean Souza, Billy Kinney, Jessie Steele. *Front row, left to right*: Devin Kamealoha Forrest, Aunty Beryl Blaich, Lahela Correa, Kauʻi Fu, Nathaniel Tin-Wong, Dominique Cordy, and Poliahu Wong. (Photo by author)

3   *Hoa 'Āina*: Grow Communities of Care

They [some newer residents] want all the perks of living in a tight community but don't want to put in any of the work. . . . If you don't teach them, they'll just stay that same disrespectful person with zero understanding of the culture.
—Tamra Martin, age thirty, Kīlauea, Kaua'i, 2016

In today's world, many communities are virtual, global, and based on interests, style, and issues—not wind, rock, and soil. Yet, places need communities, not just private landowners or government agencies, to care for them.[15] Communities of care include *hoa 'āina*, people tied to the land, whether or not they are from there, who share and fulfill obligations to a place

With a fundamental understanding of *kuleana* as common ground, communities of care can grow to include newcomers who are both expected and mentored to contribute. Clear boundaries delineating resources and who has the right to use them lead to more sustainable management, particularly when the community of users is small.[16] This book suggests possibilities for maintaining integrity even as a community grows, based on shared values, contribution, responsibility, and work.

People moving in to a place should take time to listen and learn about their new home, its history, and how their presence may perpetuate colonialism through further change or displacement. It is easy to interact only with realtors and others in the business of serving new arrivals, or within smaller communities formed only of new arrivals. Indigenous and multigenerational residents can provide a rooted core to share history, model values, and teach specific caretaking practices. New residents should join existing groups before launching new efforts, attend meetings and workdays, share food and skills. New landowners can provide jobs and affordable housing to longtime residents, share newly bought properties for community events, and invite the families who once lived there to return regularly. Recreational users can volunteer for restoration efforts or assist with resource monitoring. Students, researchers, scientists, and government agency personnel can all contribute different forms of expertise. New members can strengthen communities of care by offering diverse skill sets to contribute to efforts that nurture health of land and people.[17]

There are so many things you can do, from picking up trash to informing tourists of where to swim. Caring starts by just having the love for the land, that passion. If you see trash, pick it up. If you see bushes on the side of the road hanging over, cut them, just those simple things, simple acts.
—Lahele Correa, age twenty-one, Hā'ena, Kaua'i, 2011

4   *Konohiki*: Share Talents and Collective Ability

> *Anything that's part of our DNA and our culture, of who we are,*
> *I think we do have a responsibility to share it.*
> —Meleana Estes, age thirty-seven, Kalihiwai, Kaua'i, 2016

Connections to natural resources are connections to other people.[18] The collective process of cultivating, harvesting, preparing, and sharing food creates close ties within and among families. These ties are as important as the food or other tangible goods produced. Sharing work and food enables communities to withstand natural disasters and geographic separation, providing sustenance in lean times so that community spirit and sense of abundance endure.[19] Sharing also promotes community in which everyone has a role: some form of expertise or gift to share, whether food, child care, grant writing, or video-production skills. Abundance and community resilience rest on bringing people and their individual gifts together to take care of resources for the benefit of many.

Sharing can extend beyond an immediate place as communities of care work together in expansive networks. Networks of communities can forward local governance efforts by sharing nonprofit status, relationships with policy makers, and funding. Groups pool their unique strengths and take turns hosting one another, leading lobbying efforts, serving on leadership boards, and stepping back to have more time with family. Expanding communities of care beyond geographic boundaries to include partners engaged in similar efforts enhances local governance.[20]

> *There are many faces of community.*
> —Kau'i Fu, age thirty, Hanalei, Kaua'i, 2016

5   *Kupa 'āina*: Prioritize Community Sustenance over
      Private Leisure

> *Local people should have places they can go that tourists can't. A lot of times you'll*
> *hear, 'Where is the aloha?' Well it kind of gets sucked out of you. . . . Living here is*
> *hard, and more people coming here makes it even harder. . . . It feels like we are just*
> *here to serve visitors.*
> —Olena Molina, Kīlauea, Kaua'i, 2016

Visitors also have *kuleana* to care for the places they vacation—to treat places and their people with respect, to heed warnings and rules, to pick up marine debris along with shells, and to leave places better than they found them. Responsible visitors research permitted accommodations and tour operators that contribute to caretaking, boycott exploitative guidebooks, get by without every convenience and chain store that exists where they live, and, ultimately, return home. Just as island communities must live within the limits of their surrounding ecosystems, it is vital that visitors also respect limits. Without constraints on visitor numbers and activities, the places they frequent will be left with little beauty or sustenance to offer anyone.

In tourist-destination areas, coastal lands now provide private benefits such as recreation, leisure, and resale value, whereas historically they provided many layers of abundant community sustenance. Coastal fishing families of Halele'a watched weather and schools of fish; helped neighbors evacuate in disasters; guarded against overharvests; hosted visitors; served as lifeguards; cleaned beaches; shared phones, hoses, and showers with neighborhood kids; and lent their yards for community parties. These responsibilities now fall on fewer people, as families who sustained beach-front communities are being forced to leave, replaced by absentee owners, seasonal occupants, and a revolving stream of vacationers. When individual properties sell and longtime residents leave, quality of life diminishes for the entire surrounding community.

Allowing visitors—who often have considerably higher incomes than residents—to purchase land can tear apart communities and the places they care for.[21] Recent arrivals replace longtime families as "hosts," perpetuating their own stories and extended vacation lifestyles on the land. Local and Indigenous communities, their knowledge and many layers of relationships, become invisible or are inaccurately portrayed as existing only in the past.[22] These perceptions can perpetuate disrespectful and destructive behavior caused by unintended ignorance and the lack of guardians to prevent it.[23]

Gradual development and constant streams of visitors infringe on residents' ability to gather food, care for burials, observe changes on the landscape, orient through vistas, teach children, and visit sacred sites.[24] In tourist destinations, sources in which to restore one's spirit and feed one's family become less accessible to residents even as housing and food become more expensive. Smart phones make it increasingly difficult to regulate where people go and how they behave. Geo-tagged Instagram, Facebook, and blog posts expose once-quiet places, now backdrops for vacation fantasies and extreme adventures. Visitors push to find the last beach with no footprints, a trail off the beaten track, that hidden waterfall or online picture they aim to replicate. Unaware and unprepared, some venture to ever more inappropriate and dangerous places and then have to be rescued by residents who risk their own lives.[25] Visitors fortunate enough to enjoy a tucked away, "undiscovered" spot can reciprocate by keeping photos and directions to themselves, rather than exposing or marketing the place. Responsible tour operators, wedding officiants, and activity providers should limit their businesses to sites where commercial activities are legal, and pay a portion of their profits towards care of those sites. Guidebook, travel blog, app, and website authors have *kuleana* to work with local community groups and government to ensure that the information they provide protects both visitors and the places they visit. Co-authorship, local certifications, and other forms of partnership engage community members in sharing firsthand knowledge to enhance the accuracy of information offered about their place. In vacation destinations, challenges to place and community are common, but respectful, well-informed visitors can lessen their impacts and even contribute positively to the communities they visit.

If you want to fish, you have a place and tools, you'll learn it, relearn from the environment. If you lose a place, you lose everything associated with that place, the lessons it could teach you.
—Kauʻi Fu, Hanalei, Kauaʻi, 2016

6   *Noho Papa*: Create Careers in Guardianship

If you are from that area you tend to take care of that area.... Get a good job that can sustain you and your ʻohana within community, so that you can stay. If [young people] have more time to be in the community, then they'll have more time to grow the community in a positive way, ... rather than being tired and weathered when they come home.
—Olena Molina, Kīlauea, Kauaʻi, 2016

Long-term local-level governance requires a stable community of families who use and care for a home area continuously. Local governance falters when younger generations are unable to raise their own families in a place, whether the cause is escalating land values, destruction of a forest or fishery, or climate change in the form of warming tundra or sea level rise. In many rural communities, young people face moving away or taking jobs in destructive industries. Communities need to create jobs for younger generations to work at keeping land and people living.

On-the-ground guardians are needed more than ever as natural environments suffer from climate change, intensifying resource extraction, and increasing tourism. In addition to enforcing rules or responding to infractions like wardens or rangers, community guardians could be paid to care for places in many crucial ways: conduct ceremony, organize community, monitor resource health and use, feed, host, and teach. Careers as guardians in fields ranging from education to policy to environmental science provide opportunities to relearn and connect through everyday work and livelihood. Young people can sustain their families while actively watching, tending, and protecting the places they love.

> [We are] trying to get the kids to grow up in it, not just going on because their friends are doing it. They actually know deep in their naʻau [gut] why they're doing what they're doing. They know what's pono [just], they know what is not; and they're not just fighting for a view or a sunset. They're fighting because some place is important, culturally and deep within, familially. You are not just from that place, you are of that place. That is your ʻohana.
> —Kamealoha Forrest, Hāʻena, Kauaʻi, 2016

7   *Mau*: Nurture Future Generations of Leaders

> The most important thing is [for my daughter to] know who she is and where she comes from. The rest of the lessons are in the landscape.
> —Kauʻi Fu, Hanalei, Kauaʻi, 2016

Cross-generational transmission of skills, knowledge, and relationships with particular natural resources is crucial to the longevity of community-based caretaking and governance systems.[26] Young people must learn not just particular steps of cultural practices, but the full process. For example, in fishing, it is not enough to learn how to throw a net and catch a fish. You also have to learn how to make and patch the net, how to clean the fish, cook it, and share.[27] In addition to teaching ancestral knowledge, it is crucial to hand down understanding of contemporary legal and political systems within which governance takes place. Longtime community leaders can teach younger generations the history of past policy processes and community agreements, while transferring positive relationships with government and other partners. True perpetuation of knowledge and skill takes sustained practice, far more than can be provided through experiences at an occasional camp, or engagement in a single meeting.

Communities can grow future generations of practitioners and guardians through focused mentorship, early career internships, and ongoing teaching from youth to young adulthood. Whereas education often focuses on programs for younger children, community efforts in Haleleʻa and Koʻolau highlight the importance of educating and mentoring teenagers. Small groups of youth who are raised with their elders and mentored through community work can grow into a bridging generation of teachers and young adult leaders.[28] These individuals are prepared to apply innovative technologies to farming, fishing, and research; navigate policy; and teach children, including their own. Young adults who play this crucial bridging role in Halele'a articulated some critical conditions for their learning: healthy natural resources and the ability to feed their families from them; cultural educational programs offering hands-on experience; 'āina-based jobs; strong peer networks; and regional community exchanges with youth from other places. Cultivating young leaders today provides future generations of knowledge-keepers and caretakers.

> *I always try to bring along my cousins, nieces, nephews. I hope they see the importance of place, where they come from. They didn't get to live the life we did.*
> —Billy Kinney, Hanalei, Kaua'i, 2016

8   *Kīpuka*: Create Protected Spaces of Reconnection

> *I am looking for a place where I can clear the land, plant food, and know that my work will continue, that I won't always have to start over. I don't need to own land, don't really like that word "own," just someplace our family can live and take care of.*
> —Nathaniel Tin-Wong, age thirty-seven, Moloa'a, Kaua'i, 2016

For most Indigenous communities across the world, land and resources can never be owned, only cared for. Communities today are finding creative ways to establish protected spaces and new models of landholding based on caretaking. Wealthy property owners from outside the community can place cultural or conservation easements on lands and protect community access to conduct ceremony and restoration. Some are gifting land back to community groups whose efforts have touched them, rather than selling or passing it on to their heirs. Family and community land trusts, cooperatives, or other nonprofits hold lands for collective use and provide job opportunities in resource guardianship. Communities can prioritize creative means to protect and set aside lands, from creation of parks to stewardship agreements to legislation. There may be no plan, funds, or manpower to actively care for a place now. However, protecting lands and waters from development and degradation today gives future generations a wealth of places to return to, even if they are forced to raise their families far away.

Even small parcels of land, collectively cared for within each community, are centers of cultural and ecological replenishment.[29] These *kīpuka* provide places to gather, cultivate local and traditional foods, strengthen community ties, organize to affect policy, teach, and learn. Informal efforts to perpetuate practice and presence on the land may be as important as formal recognition of governing authority. Each *kīpuka* offers a space of communal self-determination in which to perpetuate practices and teachings of the past while envisioning and enacting the future. Lands protected as community *kīpuka* seed and nurture efforts to care for other places. Over time, these *kīpuka* grow to connect, regenerating collective *kuleana* across the landscape.[30]

> *Wherever I go, I take the aloha and things I learned here [at home] and try to be a good representative of this place, so that people can do the same in their own places.*
> —Noah Kaʻaumoana, age twenty-seven, Hāʻena, Kauaʻi, 2016

Noah Kaʻaumoana, *lawaiʻa, hale* builder, farmer, and chef, serves *awa* at a community meeting to share information and assess the first year of the Hāʻena CBSFA. (Photo by Anuhea Taniguchi)

First annual community sharing meeting on the Hāʻena CBSFA on October 22, 2017. (Photo by Anuhea Taniguchi)

Pani: Closing

These are lessons learned from my community, people old and young who have lived their lives keeping place. Here is a lei for our home. May it help you to see your own home, and your responsibilities to that place, to all the places you may visit or think to set roots, in new ways that sing the old.

> *You can write about it though. You could write something nice about this place, . . . how very productive in serving the people that used to live here, . . . the life of sharing with one another, the food, the fish, and everything about this place, and hope it brings that kind of good stuff, coming this way.*
> —Aunty Jenny Loke Lovell Pereira, age eighty-two, Papaʻa, Kauaʻi, 2007

Pīpī holo kaʻao . . . May the stories, as always, continue . . .

Epilogue

Kū Kāhili (Standard Bearers)

My son's third- and fourth-grade class from Kawaikini Public Charter School stands poised on the edge of a dune. With one word from their teacher, "*Hiki* [you can]," the kids spurt down the incline, towels flapping, bright lunch coolers flying behind. They skitter to a stop and group up just before the trees end and the beach takes over.

The wind is starting to rise off the water, flinging spray against the far boulders of Mōkōlea Point.[1] *Pōhuehue* (morning glory vines) arc toward the ocean from the shade. Their purple flowers quiver upright in the breeze, just starting to open to the sun. The bay is sloshy with white wash and current. The kids will swim in the river today. *Ohiki* sand crabs have been busy digging holes, creating swaths of sand pyramids at crab-eye level. When the kids hit the beach, they will become crabs digging for crabs. Within ten minutes the beach will be flat again.

But for now they are quiet, gathering themselves to greet the place before starting this field trip to Kāhili. [2] The ironwoods above their heads bend away from the wind, spindly branches with needles tinged orange as if rusted by salt. Above the trees, *'iwa* (great frigatebirds) are also assembling, tilting on one wing to circle above the kids.

Only a few children notice the birds. Their cries cause the others to tilt their necks back to watch the layers of *'iwa* swirling over their heads. "They are coming to meet you," says their teacher, Kumu Lei. This is their final field trip of the school year after more than sixteen visits to the various *ahupua'a* of Ko'olau. Lei tells them, "Now that you know these places, come back and take care of them."

Today they are preparing for their end-of-the-year performance before their whole school and families. Their hula honors the seaweed-picking *'ohana* of Ko'olau. The kids have come to gather *limu* to wear as lei, picking out invasive spiny seaweed (*Acanthophora spicifera*) as they go.[3] I watch them watch the ocean, moving forward on the rocks when the swells recede, then clambering backward before the break, practicing their year's lessons of moving with the sea. While most gather *limu*, a few of the kids will stay back to watch the teachers' younger children. They play in the tide pools to keep the little kids far from the snatching waves. Two of the girls will catch *a'ama* crabs by the eyes using *'ahele*, coconut snares, a technique taught to their class by an area elder. The girls will pass their catch to their classmates, who share them around like a party platter, plucking off legs and sucking the flesh, cool from the ocean.

Kawaikini Public Charter School *keiki* catching *a'ama* crabs, with their *kumu*, Lei Wann, using *'ahele*. Lilinoe Kuhaulua-Leong, Kamahina Turalde, 'Ānela Quintero, 'Ekolu Brown, Līlilehua Taylor, Ka'iulani Rivera. (Photo by author)

Before picking seaweed, the kids will kneel in the wet sand and offer their hula for the elements that inspired it. As Kumu Lei tells them, this dance to the rustling surf, rocks, and birds, the ancestors of this place, is their most important performance.

I love coming on field trips with these kids, because these places respond to their presence and voices—maybe the first time Hawaiian language has been sung over these dunes in a century. When I come with my own family, we gather trash. I have to be watchful to stop my well-meaning three-year-old from picking up the used toilet paper scattered on the sand by illegal campers. Dogs race by without a leash, frightening the kids. Unpermitted surf schools launch flotillas of boards. The mansion the County Planning Commission approved, despite the community's contested case intervention, looms above the river. Fortress rock walls buttress the spa and infinity pool, encasing the hill local fishermen once used to *kilo*. When we come as a single family, I see only problems I cannot fix. When we come as a community on one of these field trips, I leave energized to work on possibilities.

The kids are amazed when I tell them trucks used to drive all over these dunes, spinning loops on the beach and revving sand from their tires. There were piles of rubbish everywhere. About fifteen years ago, community members, including my father, started to work cleaning up the area. They hauled in boulders to place at the parking lot, blocking vehicle access onto the beach. They started monthly beach cleanups. Volunteers hauled out dumpsters and truckloads full of trash. Today the beach is serene and unmarked by tires. Beach cleanups are less frequent because there is less to pick up, though the amount of plastic washing in from the ocean on this windward beach seems to increase with every wave.

Dad used to fax the beach cleanup announcement to the radio station in time for the regular community events calendar. Each month, he whited-out the date on his printed copy and wrote in the new one through the crusted layers. He showed up to pick up trash the first Saturday of every month, even when no one else came, even into his last months of his battle with cancer. I feel close to him here.

'Iwa circling above students at Kāhili, May 2017. (Photo by author)

The kids squeal as the shadow of an ʻiwa passes above them. Birds start to settle in the trees and skim the river, drinking water. The teacher looks at my son. "Can you start the oli when you feel your classmates are ready?" As they start to chant, a mōlī (albatross), one of Dad's favorite birds, skims low in front of us, wheels, and follows Mōkōlea out to sea.

    Kiaʻi ʻia Kīlauea e Nihokū
    (Kīlauea is looked after by Nihokū)

From the top of Nihokū, also known as Crater Hill, we can see all of Haleleʻa and Koʻolau spread out below us. Kāhili Beach is nestled at the mouth of the river valley to our left. I wait for someone in the group to start our chant for this place. These young people with me are teachers, fisher men and women, paddling coaches, weavers, researchers, videographers, composers, scholars, and parents. I have watched them each grow up and taught most in community educational programs when they were just in high school. Now, we are learning together from this place. Now, they are teaching me along with my children.

Pīlaʻa's fringing reef reaches past Kāhili, with construction trailers for the Facebook retreat looking like toy trucks on the bluff. Just below to our right, Haleleʻa and Koʻolau meet at the rocks of Puʻukumu at Kalihiwai. Across that bay lies the small shoal where Aunty Annabelle spotted octopus for her grandmother. Beyond, Anini's fringing reef arcs around toward Hanalei. The far-off peak of Mount Makana marks Hāʻena. From Makana, the mountains stack back toward Waiʻaleʻale, shrouded in gray-tinged clouds. The former sugarcane fields stretch before us, now cut by dark green hedgerows around sculpted lawns. Ahead of us is Moʻokoa, the ridge shaped like a lazing lizard, stretching down toward the sea. This lizard is said to have dug out the rivers and valleys that fall off on either side of Kīlauea Plateau. Mansions that dominate the cliffs of Kauapeʻa Beach look insignificant from up here. For all the houses built up around our town, Kīlauea is still surrounded by galloping expanses of green.

The lands where we stand are part of the Kīlauea Point National Wildlife Refuge. We can gain entry to the gates only a few times per year. This past year we began seasonal observations here, sleeping overnight each equinox and solstice, to watch the sun rise and set within the refuge.[4] This is not camping but study, observing the cycles and ecological rhythms of our home even as they are changing. How do the birds circle the point at different times of year as they land with fish for their young? Where will the sun rise? Do the albatross still arrive at the start of Makahiki season?

We are learning how the top of Nihokū Crater cuts the wind, splitting clouds to pile on the mountains or billow out to sea. We think this may explain the name of the wind of Nihokū, Aopoʻomuku, which means "to cut the tops of the clouds." We are retelling the story of Pele—goddess of volcanoes and fire, simultaneous destroyer and creator of new lands. This crater is one of a series Pele dug along the island chain as she traveled from Tahiti searching for a home. Her rival sister, Namakaokahaʻi, goddess of the ocean, followed, sending crashing waves to destroy Pele's work. Here, she collapsed half of Nihokū Crater into the sea. Three *kupa ʻāina* sisters watched the battle from the crater rim. One sister teased Pele that her digging stick must be useless, since the sea goddess seemed to be winning. Angrily, Pele retorted that she would show how her stick could be used, aimed it, and turned all three women to stone. They stand today on the edge of the crater, three rock pinnacles as a reminder of the fire goddess's visit, of the connection of this place to Kīlauea on Hawaiʻi Island, where Pele now resides.

We share this story, and our learning from this place, by hosting educational groups at Nihokū. This year we taught over one hundred hula practitioners, community members, and students of all ages the stories of this place while engaging them in weeding and restoration of native plants. We have brought offerings and conducted the first Makahiki ceremonies held here in two centuries. For our children, being here is normal. They've memorized all the bird burrows. With wide eyes and a finger to their lips, they reach up for visitors' hands and introduce them to nesting *koaʻe* (tropicbirds). They kneel down in the dirt to peek into an *ʻuaʻu kani* (wedge-tailed shearwater) burrow to check if its egg has hatched.

Community members worked throughout the early 1980s to raise funds and congressional support to purchase Nihokū (Crater Hill) and Mōkōlea Point, which was set to be carved up for ten luxury homes. There were few seabirds here then, and Fish and Wildlife Service (FWS) saw little ecological value in expanding the existing Kīlauea Lighthouse Refuge. With fencing and trapping of cats and rats, the birds have since recolonized, making this place the largest concentration of native seabirds in the main Hawaiian Islands. Now, we need permission, a permit, and escort from FWS to ensure that our activities here don't harm the birds. We receive training on the birds' ecology so that we can assist in their monitoring and protection and guide groups on our own.

We are clearing weeds that have overgrown native outplantings. Community members who roamed these cliffs as kids have not been able to come here for decades. On our workdays, these families wield machetes in the hot sun so that *lauhala* trees can breathe. Their grandchildren will gather the leaves to learn to weave. Like the clouds, this is a cycle.

The hill we're standing on was too steep for sugar. Though cattle ran here, it was never bulldozed, and has remained a sheltered spot. But homes are rising up the crater. One with an entrance that reads "Sumo ʻĀina" has just been built. Grass is just starting to emerge from plugs in the graded dirt along the driveway. Survey stakes with orange streamers signal another building coming just upslope. These homes lie within the only gated community on this stretch of coast, access limited to residents. Another nearby home within the complex makes me wonder how anyone could want a view so badly that they would place their house where an entire community now sees nothing else as we drive up our town's main street. At the last Kīlauea Neighborhood Association (KNA) meeting, the owners sent their representative to ask for a letter to the island utility company requesting that poles be moved to enhance their view.

When Uncle Gary Smith—who helped lead the Kāhili Beach cleanups— called to invite my husband and I to hike up to Nihokū and Mōkōlea, our third child was six months old. I strapped her to my back and followed the men down a steep path I had not walked since my childhood. Uncle Gary, whose father was the Kīlauea Sugar Plantation manager, is a dedicated non-Hawaiian student of Hawaiian history, language, and culture, and our long-time community historian. He brought me on the hike that day to give me *kuleana*. He asked that I compose a chant for Nihokū, Mōkōlea, and

Nīhoku standing watch above Kīlauea, 1924 (Hawaii State Archives, Photo by 11th Photo Section Air Service U.S.A.).

Kiaʻi Lahela Correa, Kauʻi Fu, Olena Molina, and Leah Sausen with their *keiki,* Kaiuʻi Grady, Keʻalohi Molina, Lyla Ornellas, and Ariana Molina on the slope of Nīhokū, looking over Koʻolau and Haleleʻa. (Photo by Anuhea Taniguchi)

Kāhili, places he has spent his life working to protect. He felt the places needed to hear their names, that their stories need to be sung and shared.

The young people I stand with today, preparing to chant for this place, all carry *kuleana,* each in their own way and with grace. The hours they spend here at Nihokū and in the countless other community fund-raisers, teaching endeavors, and gatherings they run are all volunteer. Whether we grew up in this community or not, whether we are Hawaiian or not, we all carry specific *kuleana* that continue to follow and find us. As some of these young people laughingly say, "You can't outrun *kuleana.*" With perpetually increasing house prices, few in this group will ever be able to afford to buy a home

View of Halele'a from Nihokū, Ko'olau, Kaua'i, October 2017
(Photo by Anuhea Taniguchi)

in our community. Yet they strive to live in a way that leaves this place better.
When the *kuleana* they carry becomes too much, they help each other and
take turns. We are working together to prepare our own children for the
kuleana they too will carry, that they may do so naturally and with joy.

Our chant that grew from Uncle Gary's request on the hike that day describes the resilience of this community. It names the beloved places that
community members, including my own parents, have worked so hard to
protect. Whether they are held by a local land trust, the Fish and Wildlife
Service, the county, or private landowners, these places continue to shape,
nourish, and restore us. As we try to learn from, restore, and protect them.
We share these places with our *keiki* of Halele'a and Ko'olau so that they too
may love and care for their home; so that they may feel the presence of their
ancestors here when we are gone.

It feels good to stand here quietly with my children, with my teachers and
theirs, watching the clouds cycle over our home. A voice rings clear in the
soft breeze, which I can feel on the inside of my ears. We all join in to repeat
the first line: "*Kia'i 'ia Kīlauea e Nihokū.*"

Closing *Oli*

Kiaʻi ʻia Kīlauea e Nihokū

Kīlauea is guarded by the crater Nihokū, standing in the glistening sea of Makapili. At the top of Nihokū there is a space of calm despite all the changes on the surrounding landscape. The juvenile albatross of our area search, shifting from foot to foot as they get ready to fly for the first time. At Kāhili Beach, the sand heaps up into sheltering dunes in the face of stiff winds and waves that fling themselves ceaselessly toward shore. The community's endurance is seen in the return of the seabirds, soaring *ʻiwa* (great frigatebirds) and dancing *koaʻe* (tropicbirds), to once-threatened Mōkōlea Point. Mōkōlea is steadfast in the tossing ocean, moving yet rooted, like the community members whose love brings growth and aloha always. Only aloha . . .

Kiaʻi ʻia Kīlauea e Nihokū
Kū i ke kai malino aʻo Makapili
Pili nā kaikaina i ka wahine a ka lua
Lua haʻa i ka malu maiā Kekoiki
He ʻiniki ka ihona a i Kāhili
Hili hele nā maka o ka pūnua
Nuʻa ke ʻone i ke ʻalo Uhau
Hahau nā nalu poʻi mau i Kaʻiwa
ʻIwa kīkaha, puni ke ao
Ao mai nā koaʻe pili i ke ʻoni
Onipaʻa Mōkōlea haʻa i ke kai
Kaiāulu ke aloha e
A he aloha wale no e

Pilina ʻĀina: Hawaiian Terms for Some of the Many Relationships People Have with Land

Pilina (Relationship)	ʻŌlelo (Words)	Manaʻo (Meaning)	Kumu (Derivation)
Genealogical Ties (one whose ancestors are of the land)	*ke ēwe hānau o ka ʻāina*	The lineage born of the land: "A Native Hawaiian who is island-born and whose ancestors were also of the land"[1]	*ēwe* = sprout, rootlet, birthplace, family trait, or to sprout
	ʻōiwi kulāiwi	Refer to those whose ancestors bones (iwi) are planted in the land	*ʻōiwi* (native) *kulāiwi* (native land or homeland)
Stay, Dwell, Maintain Presence	*noho papa*	To dwell in one place for generations	*noho* = To live dwell, maintain space or presence in a place
Care For and Cultivate To Eat and Feed	*makaʻāinana*	One who attends the land, watches it closely	*maka* = new plant shoot, eyes of the land
	kupa ʻai au	Native	One who has eaten from the land for a long time
To Know a Place Well	*kamaʻāina* *kupa ʻāina* *hoʻokupa*	Child of the land Citizen, native To become a citizen	To be acquainted or familiar One who is well-acquainted Get to know a place well
To Protect and Care For	*hoa ʻāina*	Friend of the land	One bound to the land, companion who cares for the land.
	konohiki	Head man or woman	Lit., to invite ability

Hāʻena Community-Based Subsistence Fishing Area Rules (Haw. Admin. R §13-60.8)

1. Establish Mākua Puʻuhonua Area as a marine refuge, nursery habitat for juvenile reef fish
2. No commercial harvest
3. No fish feeding or use of devices as attractants
4. No take of empty shells using an underwater breathing apparatus other than snorkel
5. No more than twenty total living ʻopihi, pipipi, kūpeʻe, or pūpū[1] per person per day (effective November 30, 2017)
6. No more than two lobsters per day by hand-harvest only
7. No more than five urchins per species per day
8. No more than two heʻe per day, harvested only by hand or use of a stick no longer than two feet
9. Harvest seaweed by hand only
10. Allowable fishing methods:
    - up to two hook-and-lines with up to two hooks per hook-and-line
    - three-prong spear no longer than eight feet, used between 6:00 a.m. and 6:00 p.m.
    - throw net
    - paʻipaʻi net or surround gill net fishing methods, deployed from the shore or from a vessel less than fourteen feet long; at least two people must be within five feet of the net at all times
    - scoop net between 6:00 a.m. and 6:00 p.m., for no more than three marine life per day
11. Review rules in five, ten, and twenty years

Notes

Notes to Prologue

1   Oʻahu is the most populated of the seven major Hawaiian Islands, where the largest city in the islands, Honolulu, and the famous tourist destination, Waikīkī, are located.

2   *Holoholo* means to go out for a walk, sail, or stroll, to ride anywhere aimlessly (Pukui and Elbert 1986). It is a common expression used to mean fishing (see chapter 2).

3   Tapered wooden boxes allowed fishermen to look into the narrow top and through the glass viewing pane at the bottom to see underwater. Few fishers in those days had glass goggles, and there were no plastic dive masks.

4   *Puakenikeni* (*Fagraea berteriana*) is a shrub native to the South Pacific and popular for the fragrance of its flowers, used in lei making. Its Hawaiian name literally translates to "ten-cent flower," referring to a time when the flowers are said to have sold for ten cents each (Pukui and Elbert 1986).

5   Some movies filmed here include *South Pacific, Castaway Cowboy, Jurassic Park, Pirates of the Caribbean,* and, more recently and more related to the themes of this book, *The Descendants*, starring George Clooney.

Notes to Chapter 1

1   This is a paraphrase of the introduction to a community publication on the history of Hanalei, which goes on to state, "In such places, a scarcely spoken but almost tangible gladness and pride is rooted in shared experiences, in common history and in cultural values that are expressed in the daily home life, work, recreation of the residents and in the look of the locality. Communities with this 'sense of place' are becoming rare" (Blaich and Robeson 1986).

2   "Uncle" and "Aunty" are honorific terms in Hawaiian society, widely used with respect to refer to those who are not blood relatives. Except in the prologue, all the uncles and aunties in this book are related to me not by blood, but by community.

3   Hall 2017.

4   County of Kauaʻi 2016.

5   Gonschor and Beamer 2014, p. 71.
6   Hoʻoulumahiehie and Nogelmeier 2006.
7   Andrade 2008.
8   Kirch and Rallu 2007.
9   Ostrom 1990.
10   Vaughan and Ayers 2016; Higuchi 2008; Kosaki 1954.
11   Pukui and Elbert 1986.
12   Anakala Eddie Kaʻanana, personal communication, April 2003.
13   Ibid.
14   Goodyear-Kaʻōpua 2011, p. 131. Dr. Goodyear-Kaʻōpua is also a great-grandaughter
     of Papa James Kaʻōpua, whose story is shared in the prologue.
15   Stepath 2006; Vaughan and Ardoin 2014.
16   Vaughan and Ardoin 2014; DBEDT 2017.
17   County of Kauai 2017.
18   Recent news projected a 42.6 percent increase in visitor arrivals from 2017-2018 as
     airlines add new direct flights to Kauaʻi (McCracken 2017).
19   Chipper Wichman, 2010.
20   Vaughan and Caldwell 2015.
21   Johannes 2002.
22   Wilson 2009; Least Heat-Moon 1999; Simpson 2011; Kimmerer 2013.
23   Lipe 2015.
24   Moʻo is the root word for moʻokūʻauhau, geneaology (Peralto 2014), the succession
     of ancestry that lets people know who one is and where one comes from. Moʻo is
     also the root word of moʻopuna (grandchild or sucession of the spring).
25   Okri 1996, p. 21.
26   Although Hawaiian was solely an oral language, missionaries who started arriv-
     ing in the 1820s created a Hawaiian alphabet to translate the Bible and taught
     Hawaiians to read. Literacy provided a tool for Hawaiians to preserve their
     language, knowledge, and culture at the same time their population was declining
     from introduced Western diseases. Between 1834 and 1948, more than a hundred
     Hawaiian-language newspapers were published, with Hawaiians writing from
     across the pae ʻāina (archipelago) explicitly to record knowledge of cultural prac-
     tices, moʻolelo, and current events for future generations. Efforts in recent decades
     to digitize and translate these newspapers have made them more accessible,
     providing a rich resource for advancing Hawaiian scholarship (Steele 2016). See
     scholarship by Noenoe Silva, Leilani Basham, Kekeha Solis, Laiana Wong, Kauʻi Sai,
     Kamealoha Forrest, and Kahikina Silva to learn more about Hawaiian language
     newspapers and the wealth of knowledge they contain.
27   Spoken by Dr. Te Ahukaramu Charles Royal at the World Indigenous People's
     Conference on Education in 2005.
28   Some of their struggles are shared by non-Hawaiian families who have also lived
     in the area for six, seven, or more generations, many of whom descend from
     Asian migrants who came to work at area sugar plantations in the mid-nineteenth

century, and by newer migrants from California and other parts of the United States who came to Kaua'i in the 1960s-1970s and bought property that they too are now having to sell.

29 Musician Walt Keale expresses this idea very powerfully in introducing his recording of "Nā 'Aumākua," (The Ancestors). The song is a traditional chant, with lines added by Dr. Pualani Kanaka'ole Kanahele, which honors and invokes one's ancestors (Keale 2009).

Notes to Chapter 2

1   Quoted in Maly and Maly 2003, p. 868. To supplement interviews I have conducted with community members of Kaua'i's north coast over the past twenty years, I also draw on oral histories conducted by other researchers with area elders in order to provide a wider view (Maly and Maly 2003). That collection includes oral histories of elders I was unable to interview, many of whom have passed on. In many cases, I have interviewed their sons, daughters, nephews, nieces, or grandchildren, so including interviews from other sources provides multigenerational memories of place dating back over one hundred years.

2   Pukui and Elbert 1986.

3   Hawaiian genealogies, including the Kumulipo (Lili'uokalani 1897) and the story of Wākea (Sky father) and Papahānaumoku (Earth mother) (Kame'eleihiwa 1992), include all elements of 'āina, from individual islands to coral polyps to limu, fish, seabirds, and plants, as ancestors of the Hawaiian people. Relationships with the natural world were specific, varying by family, by individual species, or even by specific individuals within a species. However, all elements of nature were considered capable of interaction, part of the genealogy and community of humans. All, including physical elements such as winds, waves, and currents, demanded respect. According to Kanaka'ole Kanahele and her student, scholar Ku'ulei Higashi Kanahele, an appropriate way to refer to natural resources is akua (gods), highlighting the sacredness of all elements of the natural world (http://nomaunakea. weebly.com/ke-ki699iaka-papak363-makawalu.html).

4   Berkes et al. 2000. This worldview is common in Indigenous societies including in Europe prior to medieval times (Callicott 1994; Berkes et al. 2000).

5   Pualani Kanaka'ole Kanahele, 2016, spoken during a IUCN World Conservation Congress high-level discussion titled "Connections: Spirituality and Conservation" on September 5, 2016.

6   Oliveira 2014; Andrade 2008.

7   Aunty Pua Kanaka'ole goes on to say that ceremony is a means of establishing relationship with persons, places, and these deities, while teaching respect for all three. Fisher men and women and community members, though not trained in or making use of formal ceremony, nonetheless describe the same underlying respect and awareness of their natural surroundings as manifestations of a greater power (Kanahele et al. 2016).

8   Tommy Hashimoto, 2010.

9   Friedlander et al. 2013.

10   David and Linda Akana Sproat, 2015. Kauapeʻa Beach today is often incorrectly
     referred to as "Secret Beach."

11   Stones that influenced fish behavior are known as *kūʻula*. The first *kūʻula* came
     from the story of ʻAiʻai and his father, Kūʻula, a master fisherman of Hana, Maui
     (Nakuina, in Manu and Others 2006; Kahaulelio and Nogelmeier 2005). Kūʻula-
     kai is also the name of an *akua* (god) of fishing. *Kūʻula* stones were placed at
     key points in fishponds, to call fish into the pond, or upright along the shore
     and coastal fishing trails as shrines. Here *lawaiʻa* shared the first and best of
     their catch to give thanks to the fishing god for assistance securing it, ensure
     abundant future harvests, and lift *kapu* (protections) so that the rest of the fish
     could be eaten by people (Handy 1985, pp. 298–300). Beginning with the fall of
     the *kapu* system, when temples and images of Hawaiian *akua* were destroyed
     during Kamehameha II's reign (1819), it is likely that the practice of *kūʻula* also
     began to go underground. Some Kauaʻi fishing families describe their elders
     disposing of these family stones much later, in the 1950s and 1960s, sometimes
     by dropping them into the ocean. Others recall older family members continu-
     ing to maintain stones through the 1970s. Whether installed as shrines along
     the coast or kept within one's home, *kūʻula* required tending through certain
     observances, rituals, and offerings and were considered a source of fishing
     mana (spiritual power) and relationship with fish. The word *kūʻula* was never
     used in any of these interviews, nor was reference made to older practices
     associated with these stones, except for use of the word "shrine" in the above
     example. This practice of tending *kūʻula* is being revived by younger fishermen
     across Hawaiʻi today, including a few within Haleleʻa, some of whom learned to
     offer their catch to give thanks at fishing shrines on the island of Kahoʻolawe,
     which functions as a space for rediscovery and teaching of many cultural
     practices.

12   David Sproat, 2015.

13   Ibid.

14   Tommy Hashimoto, 2010, supported by Nancy Piʻilani, 2007.

15   Pukui, Haertig, and Lee 1973.

16   Other individual sharks were considered sinister and to be avoided, such as *niuhi*
     (man-eating sharks), many of which also appear in historic *moʻolelo* with specific
     names (Nakuina 1994).

17   In precontact stories of Hāʻena, *aliʻi wahine* (chiefesses) were said to have breastfed
     sharks on the rocks next to the channel at Kēʻē.

18   Kamealoha Forrest and SK, personal communication, 2015.

19   Maly and Maly 2003.

20   Kamealoha Forrest, personal communication, 2015.

21   SK, 2015. Another of the fisherman's nieces (KN 2017) also recalled being told of
     this incident.

22 Bobo Ham Young, 2010.
23 Valentine Ako, 2007.
24 Ibid.
25 Haunani Pacheco, 2015.
26 Linda Akana Sproat, 2015.
27 Ibid.
28 Blondie Woodward, 2015.
29 Cadiz, Peralto, and Kagawa 2015.
30 Blondie Woodward, 2015.
31 Kainani Kahaunaele, 2015.
32 Sam Meyer, 2015.
33 Llwellyn Woodward, 2015.
34 Turner 2008.

Notes to Chapter 3

1   Anderson 2013.
2   Coastal areas such as kelp beds of the Pacific Northwest were also tended by Indigenous families (Turner 2008).
3   Vaughan, Thompson, and Ayers 2016.
4   Hawai'i Kingdom Declaration of Rights 1839; Chapter III, Section 8, Laws of 1842; Civil Code of 1859, Section 387. Under Hawai'i Kingdom law, *konohiki* were allowed to designate one species for their exclusive harvest, or reserve one-third of all catch. In consultation with *maka'āinana*, they could also restrict or prohibit fishing to allow the resources time to replenish (Kosaki 1954; Higuchi 2008). The *konohiki*'s actions were not to infringe on the ability of *maka'āinana* to feed themselves and their families (Akutagawa, Williams, and Kamaka'ala 2016).
5   United States 1911, pp. 67–68.
6   Kosaki 1954.
7   By 1904, many *konohiki*, particularly those who could afford to pursue court cases to avoid condemnation, were not Hawaiian families. Damon, for instance, was a missionary descendant who purchased the *ahupua'a* of Moanalua, O'ahu, along with the nearshore *konohiki* fishing rights associated with the land (Damon v. Hawai'i, 194 US 154, 1904). Because many *konohiki* fisheries had become commercial, *ahupua'a* fishing rights were lucrative. Many court cases related to condemnation focused on how to determine a fair purchase price.
8   Kosaki 1954, p. 7.
9   HRS 2015a; Sproat 2016.
10 Kosaki 1954; Higuchi 2008; Vaughan and Ayers 2016.
11 Maly and Maly 2003.
12 Presley Wann, 2014.
13 Lei Wann, 2014.
14 Bobo Ham Young, 2010.

15   Wilma Holi, the daughter of a *konohiki* fisherman on the other side of Kaua'i, remembered visiting Halele'a to *holoholo*. Her father took her first to a particular *heiau* (temple) to leave an offering, then to visit Tai Hook's home to pay their respects before fishing (DLNR public hearing on Hā'ena CBSFA rules, public testimony, October 2014).

16   Valentine Ako, 2007.

17   Ibid.

18   Annabelle Pa Kam, 2013.

19   Kainani Kahaunaele, 2015.

20   Tommy Hashimoto, 2009.

21   In many areas of this coast, these tidal feeding patterns of fish hold despite climate change and reported declines in abundance.

22   Kalani Kahaunaele, 2015.

23   The word *imu* is also the name for an underground oven to bake food such as pig, taro, and breadfruit. While Hā'ena elders refer to *imu kai* as simply *imu,* I used the term *imu kai* here to avoid confusion for readers less familiar with the coastal practice.

24   Paulo et al. 1996.

25   Tommy Hashimoto, personal communication, 2014.

26   Kamealoha Forrest, personal communication, 2010.

27   The area, Kalei, is named for Kalei Kelau. Kelau, the last Hā'ena resident known to build *imu kai* in the ocean, was swept out to sea trying to retrieve his nets in the tidal wave of 1946 (Tommy Hashimoto and Kawika Winter, personal communication, 2014).

28   Kawika Winter, personal communication, 2014.

29   Annabelle Pa Kam, 2013.

30   Haunani Pacheco, 2015.

31   Kapeka Chandler, personal communication, 2008.

32   Jeff Chandler, 2009

33   David Sproat, Nalani Hashimoto, Keli'i Alapa'i, personal communications, 2010.

34   Today, the north shore of Kaua'i is known for its surfing, and breaks are crowded with tourists and residents. People travel and even move to the island from all over the world (including families from Brazil, California, and the East Coast of the United States) to pursue the sport of surfing both recreationally and professionally.

35   Kekoa Akana, 2014.

36   Ibid.

37   According to expert *lawai'a* Walter Paulo, different families had *kuleana* to *mālama* each of Miloli'i's four offshore *'ōpelu ko'a*, marked by triangulating landmarks on the coast. His job as a child in the 1920s and 1930s was preparing *palu* (food for the fish or bait) by grating raw pumpkin, *kalo*, or sweet potato and cooking it into porridge every day before school. After school, two people paddled a canoe out to the family *ko'a*—often he and his mother. The porridge was placed into a tiny handkerchief tied with rope so that it could be pulled to release a flurry of food

into the water. Families fed schools at the *koʻa* daily for months without harvesting. After four months, each family roasted a pig and scattered its bones into the ocean at the *koʻa* to open the fishing season. To harvest the fish, they lowered a special net over the side of the canoe, released the bait cloud inside the net and then drew the net toward the surface, tightening it to capture a portion of the school. Paulo described releasing the first fish back into the ocean to bring the school back in future harvests. After the *ʻohana* in charge of the *koʻa* took the first catch, anyone from the area could harvest at the *koʻa*. This example shows that while one family had *kuleana* to care for and create abundance at the *koʻa*, everyone benefited from this *kahu* (caretaker) family's work. *ʻŌpelu* fishing by Miloliʻi families continues to be practiced today, with community educational programs reinvigorating collective ceremonies to reopen *koʻa*, and teaching Miloliʻi children to make *palu* and catch *ōpelu* at the *koʻa*. These practices have continued largely through teachings by elders like Uncle Walter, who passed away in 2009 (Paulo et al. 1996).

38 Linda Akana Sproat, 2014.

39 Koethe et al. 2015.

40 A holdfast is a rootlike structure at the base of seaweed that allows it to hold onto a hard substrate, such as rock.

41 Annabelle Pa Kam, 2013.

42 Loke Pereira, 2007.

43 Bobo Ham Young, 2010.

44 Violet Hashimoto, in Maly and Maly 2003, p. 163.

45 Jerry Kaialoa Jr., 2015.

46 Annabelle Pa Kam, 2015.

47 Hanalei Hermosura, 2010.

48 McGregor 2007.

49 Keliʻi Alapaʻi, 2010.

50 Anabelle Pa Kam, 2015.

51 Keliʻi Alapaʻi, 2010.

52 Kawika Winter, 2010.

53 This definition for *kuleana*, as "the right to *mālama*," comes from Hawaiian legal scholar and Molokaʻi fisherwoman Malia Akutagawa. Akutagawa teaches that for decades, Native Hawaiians have had to advocate for and protect our legal rights— for example, to access ancestral harvesting areas. She points out, however, that in mobilizing to protect native rights from assault, Hawaiians have in some ways failed to emphasize the responsibilities that underpin those rights. In this way, the responsibility to access places to care for them must be protected and exercised before any right to harvest (Akutagawa, Williams, and Kamakaʻala 2016).

Notes to Chapter 4

1   From "The Life and Times of Chauncey Wightman Pa," an unpublished autobiography written in 2010.

2   According to Andrade, "*kono* means to invite, entice, induce or prompt. The term *hiki* commonly conveys that something can be done, that it is within the realm of possible" (Andrade 2008, p. 74).

3   Sharing within communities is written about extensively in other Indigenous and rural contexts. Sharing not only of goods, but of time, toward community work projects is a key aspect of social cohesion and capital (Putnam 1995). Neighboring refers to the myriad ways that community members help one another, from watching each other's children to providing food for meetings whether attending or not to lending skills for carpentry or repairs (Lukacs and Ardoin 2014). These nonmonetary contributions are essential to the closeness and resilience of a community, while also cultivating strong sense of place.

4   Peralto forthcoming 2018. *Konohiki* oversaw different levels of land division, ranging from *ahupua'a* to *moku* (districts encompassing multiple *ahupua'a*). While all were of *ali'i* rank (Kame'eleihiwa 1992), some were from the community, whereas others were appointed from outside a community by ruling island *ali'i*. *Konohiki* interfaced with *ali'i*, *akua* (gods), the *maka'āinana*, and people of the *ahupua'a*, as well as the *'āina* itself. Differences were considerable among *ahupua'a* and *moku* as to the role of *konohiki*, which also varied over time and during distinct historical time periods. Much research is currently under way to more thoroughly understand the *kuleana* of *konohiki* in different parts of Hawai'i (see the work of No'eau Peralto on Kūka'iau and Koholalele Hāmākua, and Kamealoha Forrest on Kaua'i).

5   *Ho'okupu*, as part of annual Makahiki ceremonies, displayed the fruits of community efforts to care for land and sea throughout the year by showcasing the best of the harvest. These physical offerings, along with the ceremonial prayers and *mana* (personal power and spirit of the giver), ensured future growth and abundance (Kame'eleihiwa 1992).

6   Andrade 2008, p. 75.

7   Kosaki 1954; Higuchi 2008.

8   Akutagawa, Williams, and Kamaka'ala 2016.

9   By the mid-twentieth century, some families held *konohiki* rights across multiple *ahupua'a*, with exclusive rights to fish a different, prized species in each.

10   Because some of the last registered and active *konohiki* fisheries existed in Halele'a and Ko'olau, learning from these areas can extend knowledge of the diversity and dynamism of *konohiki* systems in contemporary times (Vaughan and Ayers 2016). For an understanding of long-enduring *konohiki* fisheries in other areas, see Montgomery 2018.

11   Loke Pereira, 2007.

12   In this chapter, I use both the terms *konohiki* and "head fisherman," based on the language that interviewees used. In some areas, people said there were no *konohiki*, yet the fishermen they described played many of the same roles, overseeing

harvests and sharing the catch. In the areas where the term *konohiki* was not used, fisheries were not registered after the Organic Act, meaning that *konohiki* fishing rights were legally terminated just after 1900. In the *ahupuaʻa* where *konohiki* fishing rights were registered, they were legally recognized through statehood in 1959. This may be the reason that the word *konohiki* is applied to individuals in these registered fisheries but not to fishermen in other *ahupuaʻa*, such as Hanohano Pa, who played similar roles. Interestingly, in older stories, the *konohiki* of Hāʻena at the time of Pele's visit was a woman. Female *konohiki* definitely existed in pre-contact Hawaiʻi but are absent in the more contemporary descriptions of this role shared in my interviews.

13   *Hukilau* means to pull in the ropes tied with *lau* (leaves), once used for surround-style fishing, with nets deployed only once the shadows and movements of the *ti* leaves on the surface of the water had scared the fish close to shore. It also means "many pulling," referring to the many hands needed for these types of harvests (Vaughan and Ayers 2016).

14   Vaughan and Ayers 2016.

15   Jeremy Harris, 2015.

16   Maly and Maly 2003, p. 410.

17   Jeremy Harris, 2015.

18   Maly and Maly 2003, p. 404.

19   Linda Akana Sproat, 2015.

20   Anonymous community member, 2016.

21   Nancy Piʻilani, 2007.

22   Kealoha Saffery, 2015.

23   Chun 2009.

24   Wright 2014, p. 133.

25   Kawelo 2015.

26   Verdelle Lum, 2015.

27   Hashimoto, personal communication, 2009.

28   Kawika Winter, 2009.

29   Sam Meyer, 2015.

30   Inoa Goo Aniu, 2015.

31   Nancy Piʻilani, 2007.

32   Kaipo Chandler, 2016.

33   Kealoha Saffery, 2015.

34   Loke Pereira, 2007.

35   Ibid.

36   Annabelle Pa Kam, 2015.

37   Multiple interviewees say they will never forget the sight of Kalihiwai cleared of water for about a mile and a half out.

38   Annabelle Pa Kam, 2015.

39   Merilee Chandler, personal communication, 2015.

40   Shorty Kaona, 2015.

41   Carol Paik Goo and Nani Paik Kuehu, 2015.

42   Pa 2010.

43   Linda Akana Sproat, 2015.

44   Ibid.

45   Pa 2010.

46   Linda Akana Sproat and David Sproat, 2015.

47   During World War II, many of the reefs and beaches in Halele'a and Ko'olau were lined with barbed wire to prevent a feared Japanese invasion. In the 1946 tidal wave, some wire may have washed ashore, entangling residents also caught in the waves.

48   Annabelle Pa Kam, 2015.

49   Iniki, meaning "sharp and piercing, as wind," was a category 5 hurricane. Had it hit a more populated area, such as the island of O'ahu, experts estimated that it would have caused far more damage and loss of life (Drewes 2017). It caused six fatalities total, four of which were from Kaua'i—two on-island and two offshore (Central Pacific Hurricane Center, n.d.).

50   Nani Paik Kuehu, 2015.

51   Gary Pacheco, 2015.

52   Vaughan and Vitousek 2013.

53   During the summer, the ocean is calm, surf is flat, and schooling species migrate into the area's bays. In our study, sharing from summer surrounds did in fact provide the majority of the catch (65 percent) and an even larger portion of the sharing (over 82 percent). Individual throw net fishing provided fish for families and special occasions such as graduation parties and funerals year-round, even when the ocean was rough (Vaughan and Vitousek 2013).

54   We counted each transfer of fish, regardless of weight or number, as one "distribution." We also assumed that for every catch some was "distributed" to a fisher's own home—an assumption that is usually, though not always true. This study logged over fifty catch events, totaling 4,231 kg of fish and two hundred catch distributions (Vaughan and Vitousek 2013).

55   Kealoha Saffery, 2015.

56   Vaughan and Vitousek 2013.

57   Bobo Ham Young, 2009.

58   'Āina momona means "fat, fertile, rich, or sweet" lands. This expression refers to "abundant and healthy ecological systems in Hawai'i that contribute to community well-being" (http://kuahawaii.org/). In turn, people cultivate this abundance through active stewardship of land and sea. Thus, 'āina momona means both abundant lands and thriving people.

59   Andrade 2008; McGregor 2007.

60   Hanalei Hermosura, 2009.

61   Vaughan and Vitousek 2013.

62   Goodyear-Ka'ōpua 2011.

63   Bakunin 1882.

64   McGregor 2007, p. 17.

65   US Census Bureau 2016; DBEDT 2017.

66   Carpenter 2017.

Notes to Chapter 5

1   McGregor 2007, p. 8.

2   Vaughan 2014.

3   Some of their struggles are shared by non-Hawaiian families who have also lived in the area for six, seven, or more generations, many of whom descend from Asian migrants who came to work on area sugar plantations in the mid-nineteenth century. Newer migrants from California and other parts of the US mainland who came to Kauaʻi in the 1960s and later (some of them to surf) are also now having to sell properties they bought in Koʻolau and Haleleʻa.

4   Kuykendall 1938, p. 220; HRS 1959.

5   Garovoy 2005; the words stamped on all titles to lands issued in the Kingdom of Hawaiʻi.

6   Kameʻeleihiwa 1992.

7   Pukui and Elbert 1986.

8   Andrade 2008; Garovoy 2005.

9   Through the Māhele, all lands in Hawaiʻi were divided into crown lands (those held by the king), aliʻi lands (those held by the various chiefs), and government lands.

10 Beamer 2014.

11 Garovoy 2005; Hawaiian scholarship on the Māhele suggests Kamehameha III was strategically attempting to protect his people in the face of foreign pressure to buy land by creating an ownership system recognizable to European and American powers, yet predicated on Hawaiian values. This is one of many examples of ʻōiwi (those whose ancestors' bones are in the land) agency, the adoption of foreign tools, such as mapping, by Hawaiian monarchs, in an effort to forward their own agendas and work in the best interests of their people. However, Kamehameha's attempts were undermined in multiple ways, leading to the end result of dispossession (Beamer 2014; Kameʻeleihiwa 1992).

12 While the Māhele and the Kuleana Act created land titles, thus enabling alienation of Hawaiian lands, other factors may have been equally or more important causes of dispossession. The later Mortgage Act of 1874 actually caused families to lose their kuleana land titles by allowing predatory lenders to privately auction off borrowers' deeds without due process or judicial oversight (Stauffer 2004).

13 Kameʻeleihiwa 1992.

14 Hawaiian historian Donovan Preza argues that, though makaʻāinana were given less than 1 percent of the land in Hawaiʻi through the Kuleana Act, the intent of the Māhele was to keep reserves of land with the government for allocation to future generations rather than giving out all the land at one time. This may have been because privatization of land was an experiment, and such a new thing for the kingdom (Preza 2010).

15 Most government land grants were purchased by Hawaiians until around the time of the Bayonet Constitution of 1887, forced upon then King Kalākaua at bayonet point by a group of American businessmen and sugar barons (Preza 2010). Land

sales switched from majority Hawaiians to majority non-Hawaiians, and the size of tracts increased substantially with large sales to sugar companies. Prior to the overthrow of the monarchy in 1893, sugar companies had to lease land; after the overthrow they began to buy it outright. Other means for Hawaiian families to obtain title to larger tracts after the Māhele, besides *kuleana* awards or purchasing government land grants (Preza 2010), were organizing *hui* (collectives of area families) to buy back lands from area *konohiki* (Stauffer 2004; Andrade 2008) (see section in later part of this chapter), or leasing crown lands (Cordy 2016). Crown lands were also inalienable until the overthrow of the monarchy, when foreign settlers who had been leasing large tracts of land for sugar plantations, ranches, and other commercial ventures immediately began to buy them outright (Preza 2010).

16   There was also variation in the land awards process between *ahupua'a* and across the archipelago depending on the surveyor and area *konohiki*. Some surveyors included fallow lands, though many did not. There was no uniform standard (Cordy 2016).

17   Kame'eleihiwa 1992; Garovoy 2005; McGregor and Mackenzie 2014.

18   Gonschor and Beamer 2014.

19   Kamana Beamer pioneered use of the term "'ōiwi agency" to describe this adoption of Western tools to perpetuate native ways. He offers Kamehameha III's effort to apply Western cartography as an example, mapping all of the *ahupua'a* and land divisions of the kingdom to make visible Hawaiian governance systems on the land (Beamer 2014).

20   *Hui ku'ai 'āina* existed on other islands as well, such as the Kahana Hui in Kahana Valley on O'ahu (Stauffer 2004).

21   Andrade 2008.

22   Barrera 1984; the Hui purchased Wainiha Ahupua'a from Castle & Cooke (Dye and Dye 2010), one of the "Big Five," politically powerful agricultural (sugar plantation) companies of the time, founded by missionary descendants and other early Caucasian immigrants to the islands.

23   Andrade 2008, p. 103; in 1883 the amount of land per member increased to three hundred acres.

24   Hui Kū'ai Aina o Wainiha 1879; Andrade 2008. The *hui* movement ultimately did not survive, in large part because of a Western legal system biased against it, including one particular piece of legislation, the 1923 Partition Act, which enabled forced partitioning of Hui lands into fee simple lots, which were eventually sold (Stauffer 2004; Solis 2013).

25   The Hui became less active after 1900, as owners passed away and many families relocated (Solis 2013). "McBryde Sugar Co. actively bought shares in the Hui Ku'ai 'Aina o Wainiha," and the "Hui Ku'ai 'Aina o Wainiha was dissolved in 1947, following legal action by McBryde" (Dye and Dye 2010).

26   Andrade 2008.

27   Pukui 1983, no. 455; Hawaiians were generally easygoing and didn't order people off their lands or regard them as trespassers. When Caucasians began to own lands,

people began to be arrested for trespassing, and the lands were fenced to keep the Hawaiians out.

28 Ribot and Peluso 2003.

29 Ibid.

30 Cooper and Daws 1990.

31 Pascua et al. 2017.

32 After the second tidal wave in 1957, this road, which was previously the main highway, became a dead end, causing the communities along its length to be considered one area, all referred to as Anini.

33 Driving the coast of Haleleʻa with elders, they point out lands all along the way that used to belong to Hawaiian families. During plantation and ranching times, from the late nineteenth century to the 1970s, some families lost their lands because of inability to pay debts at local stores owned by Asian immigrant families. These stores allowed regular customers to charge items such as flour, sugar, cooking oil, lamp oil, and other staples. Store owners acquired parcels of land up and down the coast from families who could not pay off charges. "See, the way [name of store owner] inherited the property was, this family, they would charge at the store and then they didn't pay their bill. . . . Every time they charge, it gets bigger and bigger and pretty soon they took the land" (WG, 2015). In other cases, families sold land to cover debts from high medical costs incurred with serious illness of a family member (Blondie Woodward, 2015).

34 Caucasian families who moved into the area at the time found it easy to get to know and live closely among their Hawaiian and Asian neighbors. "When I first moved [to Princeville] in 1972, you had a lot of interaction. You had a number of people that worked for Princeville that were local, and you went to the same parties. You had lūʻaus, you had interaction . . . with local people" (Susan Wilson, 2015). One thirty-seven-year-old recalls how her parents rented a one-bedroom home from Wanini kupaʻāina, where she and her two younger brothers were raised. Their house was small, so they roamed the beach and trees of Wanini, and played with the five daughters of the ʻohana who rented to them and lived next door: "Sharing a piece of property with a beautiful Hawaiian family like this, that's something you just can't get these days. Growing up and knowing this is my extended family, learning all the Hawaiian values, like the openness of the house. All these people coming and celebrating and sharing food all the time. There was no 'Oh we're busy today, come back.' It was 'We're here, sure!'"(Moana McReynolds, 2016).

35 Princeville was then owned by Eagle County Development Corporation, the oil company from Colorado. Over the course of the 1980s, Eagle County developed one hotel, five thousand units of condominium, and one golf course in Princeville, with plans for additional hotels, condominiums, and a golf course spreading to pasturelands above and along the coast of Anini. Many area residents whose families had worked for the ranch and plantation now worked for the resort. Newer economic ventures such as resorts also brought in new residents drawn by construction and other jobs in these industries, who began to compete for housing

and resource access. Over time, these economic ventures became more multinational. For example, in the 1980s and 1990s, Princeville resort sold to Australian, Japanese, and, in 2005, Chinese investors. Extensive lands and enterprises such as two golf courses and two hotels, whose employees all previously worked for one company, were gradually sold off to different corporations and hundreds of individual owners. Multiple new owners are harder to influence collectively and have less sense of responsibility to either employees or the land.

36   Grading for the golf course caused heavy erosion (including one landslide in the late 1980s), which longtime area residents associate with siltation of the reef. Golf course fertilizers, pesticides, and the area landfill drain into Anini Stream, along with golf balls, which have been found by the thousands at the river mouth and on the surrounding reef. Golf balls contain a zinc core that is highly toxic to the marine environment as the balls degrade. Fishermen have observed octopus sitting on golf balls like eggs in their holes. Interviewees remember the reef as abundant through the 1960s and 1970s, with over sixty-seven different species and multiple kinds of seaweed that no longer exist in the area (likely because of water diversions and wells decreasing freshwater flow to the ocean). Discharge from cesspools of beachfront homes may also have an impact, and more recently, longtime fishermen cite increasing water temperatures, which coincide with incidences of coral disease.

37   Gary Smith, 2015; Stallone bought the property for $401,800 and sold it for $875,000 in 1998. Utah-based Castle Pines LLC, which purchased the former Stallone home for $9.75 million in 2015, sold it earlier this year for $11 million to a joint venture involving Swaying Palms Kauai Venture LP and Live Oaks San Rafael LP (Shimogawa 2017).

38   Haunani Pacheco, 2015.

39   To qualify, houses have to be occupied for more than one-third of the year and owners must be registered to vote in Kauaʻi County. "Neighbor Island Parcel" shapefile data was downloaded from Hawaiʻi's Statewide GIS Program (http:// files.hawaii.gov/dbedt/op/gis/data/NIParcels.pdf).

40   Fu, Kinney, and Vaughan 2015.

41   Total assessed value and total market value of homes were derived from the County of Kauaʻi's Real Property Tax website's mapping interface for parcel data (http://qpublic9.qpublic.net/qpmap4/map.php?county=hi_kauai&layers=parcels +ghybrid&mapmode).

42   Carol Paik and Wendell Goo, 2015.

43   DBEDT 2016; 2.8 percent of Hanalei home sales and 2.1 percent of Kauaʻi home sales as a whole went to foreign buyers, leaving only 34 percent of Hanalei sales to local buyers and only 54 percent island-wide (DBEDT 2016). It is increasingly unattainable for young local couples to buy a first home on island.

44   Coastal homes for sale searched between Kīlauea and Hanalei on Kauaʻi, using Zillow.com, as of June 2017. Extending the search to Anahola to include all of Koʻolau, the lowest price was $1.4 million on .26 acres.

45   Sotheby's International Realty (http://www.sleepinggiant.com/gallerylistings /north-shore/38).

46 These property tax rates themselves are lower than in many parts of the United States. However, in other states, such as California, property taxes stay at the same level until a home sells, then are recalculated based on surrounding value. While this creates challenges for young first-time home buyers, who have to assume higher rates, it makes it possible for families who want to continue to live in a family home without selling to do so, without facing annual increases beyond their control.

47 Hawaiian quilts are a unique style and require painstaking work, taking up to six months of stitching designs based on abstract shapes of plants and the Hawaiian flag (as a patriotic statement at the time of the overthrow), then quilting entire backdrops of stitches.

48 Lei Wann, 2016.

49 Not until 2013 did tax rates differentiate residences and houses used for vacation rentals or other businesses. Unlike California, the state of Hawai'i makes no distinction between tax rates for longtime homeowners and those of more recent buyers, forcing Hawaiian families to pay the same rates as millionaires buying second homes. Unlike other parts of the Pacific, such as Fiji or Palau, where foreigners are not allowed to buy land, because Hawai'i is ruled by the United States of America, anyone in the world can invest in property here.

50 Haunani Pacheco, 2015.

51 County of Kaua'i real property tax site (http://www.qpublic.net/hi/kauai/). For perspective, per capita income on Kaua'i was $27,441 from 2011 to 2015, with over 10 percent of the population living in poverty (DBEDT 2016).

52 Haunani Pacheco, 2016.

53 Makana Aniu Bacon, 2016.

54 CG, 2015.

55 Since 1990, visitor accommodations have proliferated in residential and agricultural areas throughout the islands, many newly constructed, some in converted garages or studios. In 2008, county government passed a transient vacation rental (TVR) ordinance to prohibit single-family dwellings being used for visitor accommodations outside of the island's five designated Visitor Destination Areas (VDAs). The ordinance provided eight months for TVRs previously operating outside of VDAs to register with the county to obtain nonconforming use certificates; 438 transient vacation rental units were registered through this process and must renew their certificates annually. Three hundred of these grandfathered properties are in Halele'a and Ko'olau. Since October of 2008, no new TVR units have been allowed outside of the VDA, and the County Planning Department actively investigates, fines, and prosecutes such operations. Some owners, many of whom do not live on Kaua'i, have resisted closure of profitable operations, resulting in costly litigation for the county. County concerns focus on negative impacts on neighborhood communities from noise and traffic congestion, as well as loss of housing and long-term rentals for local residents. Bed and breakfast operations were prohibited outside of Visitor Destination Areas through a 2016 ordinance, with twenty-five permits issued outside these areas prior to the regulation.

56   Under the Land Act of 1895, the Territorial government issued 999-year homestead leases to over 750 families between 1895 and the discontinuation of the program in 1950 (Bay and van Schaick 1994). Lessees paid no rent, but had to reside on their homestead, cultivate more than 10 percent of it, control noxious weeds, and pay taxes on the land. They could not transfer, mortgage, sell, or assign any interest in the land (Act of August 14, 1895, Act 26, [1895] Hawaii Laws Spec. Sess. 49-83). "The primary purpose of the terms of the leases appears to have been to make homesteads available to the less affluent segments of the community, to ensure that the land would be cultivated, and to protect against speculation and loss of the land" (Bay and van Schaick 1994). In 1950, the US Congress ratified Joint Resolution 12, in which the Territorial Legislature found "that it was the policy of the United States to grant fee simple patents to homesteaders, [and] that restraints on alienation were looked upon with disfavor" (Territory of Hawaii, Laws of the Territory of Hawaii, 1949, Joint Resolution 12). Joint Resolution 12 ended the Homestead Lease program and provided lessees with an opportunity to purchase their properties at fair market value and convert the leases to fee simple. Of the 750 leases issued, only sixty-five to seventy families purchased fee simple ownership. In 1994, an estimated fifty-one homestead leases, including the one described here (one of eighteen on the island of Kaua'i) remained (Bay and van Schaick 1994). The status of the other fifty remaining leases, as well as over six hundred that do not appear to remain yet were not purchased in fee simple by their holders, require further research.

57   Act of Sept. 1, 1950, Chapter 833, 64 Stat. 572, as amended; Act of Aug. 23, 1954, Chapter 824, 68 Stat. 764.

58   Bay and van Schaick, 1994. The Wanini property was converted to a month-to-month lease in 1982. More research is required to determine both the number of 999-year leases currently held by the state and when they were converted.

59   Haunani Pacheco, 2015; County of Kaua'i real property tax site (http://www.qpublic.net/hi/kauai/).

60   The family received help from the Native Hawaiian Legal Corporation (NHLC), a nonprofit legal organization dedicated to representing Native Hawaiian families in land issues. Demand for NHLC services exceeds the small organization's capacity to serve, and in most cases, families still have to pay a fee (former NHLC lawyer, personal communication, 2016).

61   Haunani Pacheco, 2015.

62   The Uniform Partition of Heirs Property Act was adopted on July 12, 2016 (effective January 1, 2017), to require stricter noticing requirements and allow heirs of *kuleana* lands the first right of refusal in actions for partition. However, Zuckerberg filed December 30, 2016, just before the new law went into effect.

63   Garovoy 2005; "Contemporary sources of law, including the Hawai'i Revised Statutes, the Hawai'i State Constitution, and case law interpreting these laws protect six distinct rights attached to the *kuleana* and/or Native Hawaiians with ancestral connections to the *kuleana*. These rights are: (1) reasonable access to the

land-locked *kuleana* from major thoroughfares; (2) agricultural uses, such as taro cultivation; (3) traditional gathering rights in and around the *ahupuaʻa*; (4) a house lot not larger than 1/4 acre; (5) sufficient water for drinking and irrigation from nearby streams, including traditionally established waterways such as *ʻauwai*; and (6) fishing rights in the *kunalu* (the coastal region extending from beach to reef). *Kuleana* rights are often associated with a Native Hawaiian ancestral connection to specific lands, but in fact, these rights can run with the *kuleana* land itself, where the courts and legislature have not explicitly stated otherwise" (Garovoy 2005, p. 530).

64 Sproat 2010.

65 Former Kauaʻi County Planning Director, Keith Nitta, 2015.

66 A longtime Kauaʻi real estate attorney and realtor team, operating since the Kīlauea sugar plantation closed in the 1970s to partition and sell former sugar lands, developed this approach of acquiring *kuleana* and moving them to increase their value. The example above from Anini is one of the first applications of this approach, along with movement of multiple taro patch parcels from the floor to the ridgeline of nearby Kalihiwai Valley. The move shifted these lots out of the flood zone and procured them expansive views. The realtors then sold them.

67 Shimogawa 2014; brochure for "The Hanalei Club," by The Resort Group, Discovery Land Company, and Reignwood.

68 Discovery Land Company boasts sales of over $2.4 billion since 2000 on one of its Hawaiʻi properties, located on the Kohala Coast of the Big Island (http://discoverylandco.com/). Since 1994, the company has been developing luxury residential private clubs and resorts in locations across the continental United States and throughout the world, including Makena, Maui; Big Sky, Montana; Los Cabos, Mexico; Great Guana Cay, Bahamas; and more. Discovery Lands was set to develop the Hanalei Golf and Beach Club at Haleleʻa with owner, The Resort Group, which owned all the former Princeville Ranch lands, and funder Reignwood International, a Chinese luxury firm that also owns Red Bull. The owner of The Resort Group, Jeff Stone, also developed Aulani, the Disney Resort on Oʻahu. In the fall of 2015, Stone announced that Discovery Lands would no longer be involved with the Princeville project, whose future has been uncertain ever since. As of 2017, it appears that The Resort Group has a new name associated with operations and assets in the Princeville area, PRW Princeville Development Company LLC (PRW Princeville). Multiple subsidiary holding companies associated with Jeff Stone, each with the same Honolulu address, hold land and other assets in the Princeville area, including former *kuleana* lands.

69 Brochure for "The Hanalei Club," by The Resort Group, Discovery Land Company, and Reignwood International.

70 The only non-*kuleana* lot at the end of Anini road was recently bought in the fall of 2016 by another subsidiary of these corporations, Anini Beach Hale LLC. All these corporations and subsidiaries have the same downtown Honolulu address as The Resort Group, owned by Jeff Stone.

71 Haunani Pacheco, 2015.

72 Shimogawa 2016.

73 County of Kaua'i real property tax site (http://www.qpublic.net/hi/kauai/).

74 Gomes 2017.

75 Ibid.

76 In personal communication, heirs who did not speak out publicly also expressed their intention to keep their shares of the land for future generations of their family (2017).

77 Wealthy landowners often have little contact with communities in which they buy land. Many, particularly high profile individuals, interact mainly with realtors, attorneys, and individuals (most of whom have also relocated to Hawai'i) who make a living managing land and building homes for such clientele.

78 Zuckerberg 2017.

79 Imagine organizing all of the descendants of your great-great-great-great-great-great-great-grandfather to make a collective decision.

80 Kainani Kahaunaele, 2015.

81 Anonymous Wanini community member, personal communication, 2016.

82 Kainani Kahaunaele, 2015.

83 Haunani Pacheco, 2015.

84 Llwellyn Woodward, 2015.

85 Jamie Pānui-Shigeta, personal communication, 2016.

86 The changes and struggles facing Halele'a and Ko'olau families are common to communities worldwide that experience ongoing settler colonialism. In settler colonialism, settlers bring a new form of government, used to establish control of land and push forward expansion (Alfred and Corntassel 2005). Settler colonialism often begins with establishment of agricultural operations, but enables a population to be expanded by continuing immigration at the expense of native lands and livelihoods (Wolfe 2006, p. 395). Settlers and their culture saturate lands, assimilate native lands and cultures, or erase them and replace them with their own (Veracini 2011). In Hawai'i, many settlers are adopting rather than erasing native cultural practices; however, their actions on the land can still hinder native residents from perpetuating these practices themselves.

87 Pukui 1983, no. 1978.

88 Inoa Goo Aniu, 2015.

89 Haunani Pacheco, 2015

90 DBEDT 2017.

91 The Hawai'i Revealed series of guidebooks, including *The Ultimate Kaua'i Guidebook* by Andrew Doughty, directs tourists to particularly inappropriate and treacherous places. Editions of the guidebook are updated about every two years, each including "new hidden gems" (Kazmirazck 2015, Doughty 2014). With each updated publication, residents on Kaua'i have observed influxes of tourists seeking dangerous places with made-up names like "champagne pools." Past listings of north shore tide pools have correlated with spikes in visitor drownings at these remote, un-lifeguarded locations. While the latest edition has removed one such

dangerous spot, it has added an even more treacherous and difficult-to-access set of tide pools that most area residents would never try to reach even in the best of conditions. In one article, Doughty is quoted as saying, "Ultimately the books and the apps are an expression of what I think about Hawai'i. . . . Once in a while if you are really lucky in life, you find the place where you belong. And I found it and I knew this is what I wanted to do with my life" (quoted in Kazmirzack 2015). In 2011, state lawmakers proposed legislation (House Bill 548 HD 3) to hold guide-book authors and publishers liable for deaths or injuries at spots they recommend.

92 Multiple Kaua'i guidebooks list beaches as excellent for year-round swimming and snorkeling, when, in reality, those beaches can be hazardous in winter (Vaughan and Ardoin 2013), which is the busy season for travelers.

93 Alayvilla 2016. Also see Blay 2011 for an insightful analysis of visitor drownings on Kaua'i.

94 There were eleven drownings in Anini from 1990 to 2008 and three in 2016 (Blay 2011).

95 ku'ualoha ho'omanawanui, 2015.

96 Kainani Kahaunaele, 2015.

97 Kamealoha Forrest, 2016.

98 An expression meaning to catch nothing but foam, or "white wash."

99 Carol Paik and Wendell Goo, 2015.

100 Makana Martin, 2009.

101 Keli'i Alapa'i, Bobo Ham Young, 2009.

102 Many of today's fifty- to seventy-year-old fisher men and women say they were discouraged from activities viewed as playing in the food source. Surfing, though a Native Hawaiian pastime, was not common in breakers of Halele'a until the 1960s, and fishermen recall hiding their boards in the bushes when they first tried the sport, lest they be scolded by older relatives.

103 In one *ahupua'a*, a well-known Hollywood actor moved the coastal trail through his property, which previously was accessed by fishing families to approach the coast along a bluff, checking ocean conditions and spotting schools from above. To maintain his privacy and avoid paparazzi, he moved the trail to the edge of his property and fenced it, so that *lawai'a* now approach the same fishing spots through a dark chute, following a stream along the back of the ridge with no view of the coast.

104 Verdelle Lum, personal communication, 2015.

105 Corntassel 2012.

106 Pukui 1983, no. 2336; land was given to people by the chiefs. Should members of the family go elsewhere, the one who dwelled on the land was considered the owner. A returning family member was always welcome, but the one who tilled the soil was recognized as holding the ownership.

107 Inoa Goo Aniu, 2015.

108 Haunani Pacheco, 2016.

109 Verdelle Lum, 2015.

110 Kainani Kahaunaele, 2015.

111 Pukui and Elbert 1986; literally one who has become familiar through eating from the land for a long time.

112 Keli'i Alapa'i, 2009.

113 Hanalei Hermosura and Keli'i Alapa'i, 2009; Presley Wann and Lei Wann, 2016.

114 Nadine and Blondie Woodward, 2015.

115 Vaughan and Vitousek 2013.

116 Pukui and Elbert 1986, p. 43.

117 Nani Paik Kuehu, 2015.

118 They were married in 1973 across the street from their family homes, under the big *kamani* tree at the Kalihikai county beach park. Their wedding photos show Uncle Blondie in a powder blue tuxedo with curly red sideburns, Aunty Nadine in a lace bibbed *mu'umu'u*. There were no pipe-frame pop-up tents in those days, so their family used coconut fronds and bamboo to construct a large shelter extending out from the park pavilion.

119 Kainani Kahaunaele, 2015.

120 Words of a beloved song describe love born of getting to know all aspects of someone's character.

121 kuualoha ho'omanawanui, 2015.

122 Ibid.

123 Kainani Kahaunaele, 2015.

124 Kamealoha Forrest, 2016.

125 Kainani Kahaunaele, 2015.

126 Physical trips are becoming more difficult as interisland airfares increase, from $45 one way in 1997 to prices of $300 round-trip in the summer of 2017. Prices are higher on peak vacation times such as school breaks, and Hawai'i residents pay the same fares as tourists.

127 Maly and Maly 2003, p. 706.

128 Pukui and Ebert 1986, p. 224.

129 Maluhia Fereira and Makaleka Fereira, personal communication, 2016.

130 Nani Paik Kuehu, 2015.

131 Andrade 2008, p. 126.

132 Beach hibiscus (*Hibiscus tiliaceus*) or *hau*, as it's known in Hawai'i, is a sprawling, flowering tree in the mallow family, believed to be a naturalized Polynesian introduction. It tends to grow rapidly and persistently, intertwining its roots and creating dense thickets that, when left unchecked, create a damming effect that negatively impacts Hawaiian stream ecosystems (Fitzsimmons et al. 2005).

133 Stepath 2006.

134 Atta Forrest, 2010.

135 Kamehameha Schools holds 375,000 acres of ancestral land across Hawai'i, endowed in trust by Princess Bernice Pauahi Bishop to educate Hawaiian youth and safeguard the perpetuation of Hawaiian language and culture.

136 Blaich 2003.

137 Ibid.

138 In 2016, Waipā Foundation served 3,700 learners, with approximately another 5,000 individuals attending festivals, farmers markets, and other one-time events on-site.

139 www.waipafoundation.org.

140 Mamaril, Cox, and Vaughan 2017.

141 Kainani Kahaunaele, 2015.

Notes to Chapter 6

1   McGregor 2007.

2   Kameʻeleihiwa 1992.

3   Menzies 2007.

4   Hāʻena is often referred to in chants and *moʻolelo* as *Hāʻena pili i ke kai*, Hāʻena close to the sea. This name describes the proximity of mountains to the sea, and area families' close connection to their coast.

5   Institutions are "structures or mechanisms of social order that govern individuals' behavior within a community" (Ostrom 1990). Institutions can include organizations, rules, informal agreements, leadership, and community ties.

6   Shomura 1987; McClenachan and Kittinger 2012; Friedlander and Rodgers 2008.

7   State of Hawaiʻi 2017.

8   Schemmel and Friedlander 2017.

9   For example, Kauaʻi Island has only fourteen DOCARE officers (Azambuja 2011) to serve the entire island, from the mountains up to three miles offshore, none of whom work at night. Interviews conducted with fourteen Hāʻena *kūpuna* in 2003 documented their concern for the declining health of nearshore fisheries (Maly and Maly 2003). The *kūpuna* suggested many causes, but felt all could be traced to lack of traditional, local-level management.

10   HRS 2015b.

11   Higuchi 2008.

12   Vaughan and Caldwell 2015.

13   Poepoe, Bartram, and Friedlander 2007.

14   By 2012, DAR had two staff positions focused on working with CBSFA and other Hawaiʻi fishing communities: one planner and one coordinator funded by a local nonprofit.

15   The group included longtime advocates for Hāʻena, including two nonprofit directors, *kūpuna*, and fishermen, among them Jeff Chandler, featured in the introduction of this book.

16   Ayers and Kittinger 2014.

17   Vaughan and Caldwell 2015.

18   At the time, the nonprofit was known as Hawaiʻi Community Stewardship Network (HCSN), established in 2003. The group reorganized in 2012 under the name Kuaʻāina Ulu ʻAuamo (KUA). As explained on their website, "KUA means back. Like a backbone that connects and supports. *Kuaʻāina* are the grassroots,

rural peoples of Hawai'i *nei*. *Ulu* means to grow. *'Auamo* is the carrying stick held on multiple shoulders of laborers who shared the burden of carrying something of great weight forward. By taking up the *'auamo*, our *kua'āina* communities share the sacred responsibility, or *kuleana*, to better Hawai'i." KUA "builds local capacity for community-based management of natural and cultural resources, supporting community-based organizations in identifying their own resource management goals and developing the expertise, knowledge, and skills necessary to accomplish these goals." The organization works only with communities that request their help. KUA runs three networks—one of forty-plus fishponds and their caretakers, one of *limu* (seaweed) practitioners, and E Alu Pū, which represents over thirty fishing communities. Each of these networks meets annually (www.kuahawaii.org).

19   Overfishing, while a threat, actually was less worrisome to community members than other factors such as high-volume recreational use (through sunscreen, walking on the reef, and disruption of the feeding and spawning patterns of marine species), proliferation of cesspools and septic systems along the coastline, decrease in the quantity of fresh water entering the ocean, and global environmental changes such as increases in ocean temperature.

20   The committee also established a vision and three goals for community fisheries management: (1) increase resource health by addressing ecosystem-based threats; (2) reduce user conflicts and impacts to subsistence fishermen; and (3) perpetuate Hawaiian cultural resource management practices.

21   Students included myself, pursuing my doctorate, and four undergraduate research assistants.

22   *Kupa 'āina*, the native families of a place, actually translates to "people familiar with a place" or, as described in chapter 5, those who have eaten from the land for many generations. In Hā'ena these are the twenty families who have ancestors in the Hui Ku'ai 'Āina o Hā'ena, who trace their presence on the land to before the Māhele of 1846–1855.

23   Vaughan and Caldwell 2015; during the writing of this book, two communities have been actively engaged in the CBSFA process: Ka'ūpūlehu on Hawai'i Island applied for CBSFA designation in 2007, and secured DLNR approval of their rules in 2015. Mo'omomi on Moloka'i, which piloted the CBSFA program in 1995, submitted its rules for the public scoping process in early 2017. Twenty other communities, half of whom submitted for designation a decade ago, are still waiting.

24   A joint resolution adopted in November of 1993, one hundred years after the overthrow, and signed by President Clinton, states, "*Whereas*, in a message to Congress on December 18, 1893, President Grover Cleveland reporting fully and accurately on the illegal acts of the conspirators, described such acts as an 'act of war, committed with the participation of a diplomatic representative of the United States and without authority of Congress,' and acknowledged that by such acts the government of a peaceful and friendly people was overthrown; *Whereas*, President Cleveland further concluded that a 'substantial wrong has thus been done which a due regard for our national character as well as the rights of the injured people requires we should endeavor to repair' and called for the restoration of the Hawaiian

monarchy" (United States Public Law 103-150. 103d Congress Joint Resolution 19 Nov. 23, 1993, "To acknowledge the 100th anniversary of the January 17, 1893 overthrow of the Kingdom of Hawai'i, and to offer an apology to Native Hawaiians on behalf of the United States for the overthrow of the Kingdom of Hawai'i").

25 Kame'eleihiwa 1992; Preza 2010; Beamer 2014.

26 There are roughly ten sovereignty organizations in Hawai'i, which advocate different models of sovereignty, including nation-within-a-nation status similar to many Native American tribes, as well as complete independence from the United States.

27 These conflicts were heightened by other state processes taking place within Hā'ena at the same time, including a thirty-year community effort to plan for local management of the area state park. Some community members opposed plan proposals to limit the number of park users as a threat to their ability to access and enjoy the area.

28 Hanalei Hermosura, 2009.

29 Drug addiction also affects area natural resources, as addicts harvest at all hours to catch species to eat or sell to fund their habit, with no regard for whether those species are legal, or for how they harvest. In one example, a young addicted fisherman placed a wire fence across a river and dropped in a charged car battery, so that all of the 'o'opu were shocked into the fence. Another young fisherman shared that when certain long-time area addicts or dealers are let out of prison, he sees the impact on certain species immediately.

30 Presley Wann, 2016.

31 Tommy Hashimoto, 2010.

32 DAR staff member, 2011.

33 Keli'i Alapa'i, 2010.

34 Chipper Wichman, 2011.

35 Petra MacGowan, former DAR staffer, 2011.

36 Taylor Camp, begun in 1969, was a hippie settlement on coastal lands in Hā'ena owned by Howard Taylor, brother of famous actress Elizabeth. Taylor encouraged people seeking an alternative lifestyle and refuge from protests and violence to move from the continental United States to live off the land on his parcel. The state condemned the land in 1973 and later closed down the settlement, which had 120 residents at its height. The former Taylor Camp lands are near the Hui taro patches within the Hā'ena State Park. Many former residents of Taylor Camp continue to live on the north shore of Kaua'i.

37 Vaughan and Caldwell 2015.

38 Jacob Maka, community meeting attendee, 2009.

39 DAR staff member, 2011.

40 Petra MacGowan, former DAR staffer, 2011.

41 Kawika Winter, 2011.

42 Ibid.

43 Presley Wann, 2016.

44 Scott Robeson, testimony in public hearing, 2014. This book focuses on the experiences of longtime, cross-generational families, particularly those indigenous to

Halele'a and Ko'olau. It is beyond the scope of this book to capture the experiences of the many non-Hawaiian community members who have moved to the area and grown into important contributors, advocates, and caretakers. For example, Scott Robeson, quoted here, took it upon himself, in his lifetime, to regularly clean both sides of the historic eight-mile stretch of road from Hanalei to Hāʻena, all alone. This road, with its single-lane bridges, defines the community in many ways, requiring drivers to slow down. Community members, including Scott's wife, have worked for decades to ensure that the road is not widened and that bridges are not replaced with higher capacity ones, in order to keep the community's rural character. Scott increased safety and visibility for drivers through his efforts of cutting back grass and weeds along the winding road. Until he passed away in 2016, Scott was a common sight along the highway, a one-man operation with multiple Weedwhackers and tools hanging off his truck, wearing a fluorescent green long-sleeve shirt, orange reflector signs, and protective eye gear. He worked for no pay and trained local youth, most of whom were Hawaiians from the Wainiha and Hāʻena community, for landscaping jobs.

45 Carl Imparato, personal communication, 2014.

46 Vaughan and Caldwell 2015.

47 Hāʻena worked with DOBOR extensively in rule development, and the agency submitted a formal letter in support of the Hāʻena rules package. However, after not acting on the rules submission for a year, DLNR removed all recreational rules not directly related to fishing, saying a separate rules package would need to be promulgated through DOBOR. The community rules related to recreation have yet to enter the formal approval process through DOBOR to become law. The community did succeed in passing the *puʻuhonua* (protected closure area) for the Mākua lagoon and hatchery. To convince DAR to keep this rule, community leaders worked with researchers at the University of Hawaiʻi and KUA to produce (1) documentation that elders viewed the area as a hatchery protected in traditional practice; (2) a literature review of scientific evidence that recreational uses can impact fish populations; (3) multiple counts and studies of human use and fishing in the area; and (4) independent research by marine scientists who conducted 110 transects to show that the lagoon is in fact a hatchery.

48 Holmes 2010; Vaughan, Thomson, and Ayers 2016.

49 Kittinger, Ayers, and Prahler 2012.

50 Jokiel et al. 2011; Higuchi 2008; in other parts of the world today, local-level governance is much more direct and flexible. In parts of New Zealand, the head fisherman of an area determines when it is time to rest a species, or particular area, then places a sign on the beach to proclaim the *kapu* (E Alu Pū Global Gathering, personal communication, 2016).

51 Though rules could be written to sunset automatically after a set number of years, community members feared this would mean termination of needed rules. Fisheries committee members also initially proposed slot limits, in which fish can be caught only within a certain size range, such as six inches to one foot. Slot limits allow baby fish to grow bigger, while leaving the largest individuals, which produce

more eggs. DAR discouraged slot limits as too complicated to enforce, arguing
also that the community would need separate scientific studies of every species to
justify proposed legal and illegal catch sizes.

52 Vaughan, Thompson, and Ayers 2016.

53 Chipper Wichman, 2011.

54 DAR staff also discouraged a ban on commercial activities allowing only subsistence harvest. Their concern was that such a ban would be impossible to enforce,
as officers would have to apprehend fishers in the act of harvesting then selling the
same catch. However, the ban was constitutional, because it precluded use based on
particular types of activity, not identity. The Hāʻena community chose to leave the
ban in the rules over the objections of state personnel.

55 Protecting these *limu* beds for elders was historically practiced in Hāʻena, as
discussed in chapter 3. Reserving more accessible gathering areas for elders was
practiced in other areas of Hawaiʻi as well (McGregor 2006, p. 16).

56 Vaughan and Ardoin 2014; Stepath 2006.

57 Chipper Wichman, 2010.

58 The second Hawaiʻi community to attain CBSFA designation, after Hāʻena, was
Kaʻūpūlehu on the Island of Hawaiʻi. Their rules package is a ten-year ban on harvest of all species in their area, including by community members, to allow it time
to rejuvenate.

59 Jeff Chandler, personal communication, 2010.

60 Alan Friedlander, UH Mānoa Department of Marine Biology, and Fisheries
Ecology Lab, personal communication, 2011.

61 Debbie Gowensmith, 2009.

62 Two commercial fishermen, one a paid lobbyist for the Western Pacific Fishery
Council, a federal advisory group for commercial fishermen, proposed allowing
continued commercial harvest of invasive fish species within Hāʻena's no-take zone
to control their populations. Allowing entry to fishing boats compromises enforcement, as vessels could always claim they are targeting invasive species. However,
after ten years of meetings and input, twenty rules drafts, and over six hundred
testimonies in support of the existing language, the Board of Land and Natural
Resources added a last-minute provision allowing permits for take of invasive
species.

63 Pukui and Elbert 1986.

64 Former head of DAR, 2011.

65 Debbie Gowensmith, 2009.

66 Peralto forthcoming 2018.

67 Debbie Gowensmith, 2009.

68 Fishermen opposed to the rules argued that Hawaiian rights protect fishing
anywhere in Hawaiʻi using any gear. This position is not supported by the history
of *konohiki* fisheries in Hawaiʻi. Legal documents, including the first written
constitutions of Hawaiʻi, protected rights to harvest in the nearshore fishery of
the *ahupuaʻa* where one resides (Kosaki 1954; Higuchi 2008). More contemporary
legal opinions on Native gathering rights on land, as well as in the sea, extend

protections only to gathering in areas where one's family traditionally harvested, even if outside one's own *ahupua'a* (Akutagawa 2016).

69 *Mahalo* once more to Hiʻilei Kawelo of Kaʻalaea and Heʻeia Oʻahu, fisherwoman, *mahiʻai iʻa* (fish farmer), hero, and friend.

70 Keliʻi Alapaʻi, personal communication, 2016.

71 Unlike the word *kilo* (to watch closely, observe, forecast, a seer, or reader of omens), *kiaʻi* includes the idea of acting on that which is observed to protect and guard a place.

72 In the Hawaiian Kingdom, for example, *konohiki* based local-level water allocations on how much labor farming families contributed to collective maintenance of irrigation ditches. Those who worked hardest were guaranteed the most water (Tong 2014).

Notes to Chapter 7

1 These provisions are offered in the tradition of past scholarship in the field of community-based and collaborative resource management (Ostrom et al. 1999; Menzies 2007; Cox , Arnold, and Tomás 2010).

2 What I am describing is quite different from commercial tours or retreats offering activities like yoga on the beach, meditation at secret waterfalls, or swims with pods of dolphins. Such ventures, increasingly common in visitor destinations such as Hawaiʻi, seek to tap into the spirit and *mana* (spiritual power) of places for personal enrichment and commercial profit, rather than to guide and inspire the work of caring for these places. Few of these ventures return monetary support, labor (to clean up litter and so on), or anything to the places they profit from taking people to visit.

3 Kanahele et al. 2016.

4 Kimmerer 2013, p. 37.

5 Ceremony does not have to be religious or affiliated with any church.

6 Tipa and Panelli 2009; Turner and Spalding 2013; Magnusson and Shaw 2002.

7 Tipa and Panelli 2009.

8 Adaptive management is a natural resource management approach for dealing with uncertainty and complexity by emphasizing "learning-by-doing" and the ability to readjust based on new insights (Berkes 2015).

9 Armitage et al. 2011.

10 Diver 2012; Ayers and Kittinger 2014.

11 Richmond 2014; Panelli and Tipa 2009; Turner 2008.

12 Kimmerer 2013; Turner, Ignace, and Ignace 2000.

13 Akutagawa forthcoming 2018.

14 Cinner and Aswani 2007; Tipa and Welch 2006.

15 Ostrom 1990; Pinkerton and Davis 2015.

16 Ostrom 1990.

17 Agrawal and Gibson 1999; Berkes 2004.

18 Richmond 2014; Panelli and Tipa 2009; Turner 2008.

19 Sol Kahoʻohalahala, personal communication, 2016.

20 Richmond 2014.

21 Koethe 2017.

22 Kaomea 2014; Raibmon 2005; DeLoria 2006.

23 Schewe et al. 2012.

24 Ibid.

25 Between July 2012 and July 2015, there were 147 visitor deaths in Hawaiʻi and thousands more rescues (Civil Beat Editorial Board 2016); between 2007 and 2016, about 323 visitor deaths in Hawaii were caused by unintentional drowning (https://cdn.relaymedia.com/amp/www.civilbeat.org/2017/08 /hawaii-tourists-are-drowning-at-nine-times-the-rate-of-locals/).

26 Ostrom 2005; Richmond 2014; Goodyear-Kaʻōpua 2012, Turner et al. 2000.

27 *Mahalo* to Uncle Tommy Hashimoto of Hāʻena for sharing this reminder with young people in his community, during a fishing camp held in the summer of 2016.

28 Research on collaborative management focuses on the crucial role of bridging organizations, nonprofits, or third party groups (Berkes 2009; Vaughan and Caldwell 2015). In community governance, bridging generations, small groups of youth, raised with their elders and mentored in community processes to grow into young community leaders, also play an important role.

29 Peralto forthcoming 2018; Goodyear-Kaʻōpua 2013; Corntassel 2012.

30 Peralto forthcoming 2018; Corntassel 2012.

Notes to Epilogue

1 A train named Pilaʻa once carried bags of sugar from the mill in Kīlauea to the end of Mōkōlea Point, where a weir loaded it onto waiting ships. In the winds and churning sea, the 125-pound bags sometimes dropped, spewing their sweet contents into the ocean and attracting large sharks.

2 *Kāhili* are feather standards that signify *aliʻi*. Those who carry the *kāhili* walk ahead to clear space for ceremony or for those with great mana. *Kāhili* are often made with seabird feathers. This practice is returning as these birds are protected and as our people return as caretakers to Papahānaumokuākea, where bird populations and the feathers they shed are plentiful. Papahānaumokuākea is the largest marine protected area in the world, encompassing the northwest Hawaiian islands.

3 Lei of *limu kala* (*Sargassum echinocarpum*) are used in ceremonies to drive away sickness and to obtain forgiveness (*kala*). Those needing to unburden themselves of some past wrong are advised to make a lei of this *limu* and walk into the ocean until the surges carry the lei away.

4 The summer solstice, when the sun travels its most northern path, is Ke Ao Polohiwa a Kāne. The sun disappears to the south for the winter solstice, Ke Ao Polohiwa a Kanaloa. The spring and fall equinoxes, during which the sun travels its longest path across the center of the sky, are Ka Piko o Wākea, a time to rebalance. Each period is affiliated with seasonal indicators and characteristics of a different Hawaiian god, and each carries different *kuleana* for people. *Mahalo* to Kumu Kalei

Nuʻuhiwa, along with her Kauaʻi *haumāna* (students) for guiding our rediscovery and observation of Hawaiian lunar and seasonal cycles of marking time.

Note to Appendix A

1   Pukui 1976, no. 1691. *Mahalo* to Dr. Kawika Winter for first calling my attention to these many Hawaiian words for different connections to land and the relationships and *kuleana* these words teach (Kawika Winter, personal communication, summer 2009).

Note to Appendix B

1   These include various small marine mollusks, such as *Nerita polita, Theodoxus neglectus, Cellana talcosa, C. sandwicensis,* and *C. exarata.*

Glossary

'ā: Red-footed booby bird (*Sula sula rubripes*)

āholehole: An endemic fish (Hawaiian flagtail, *Kuhlia sandvicensis*) found in both fresh and salt water

ahupua'a: Land division usually extending from the uplands to the sea, so called because the boundary was marked by a heap (*ahu*) of stones surmounted by an image of a pig (*pua'a*), or where a pig could be offered as tax to the chief

'āina: Land, earth; *Lit.*, that which feeds

'ako: To cut, shear, clip, trim, as hair; to break or pluck, as flowers

aku: Bonito, skipjack (*Katsuwonus pelamis*), an important food fish

akua: God, goddess, spirit, divine

akule: Bigeye or goggle-eyed scad fish (*Selar crumenophthalmus*)

ali'i: Chief, chiefess, ruler, monarch, king, queen

aloha: Love, affection, compassion, mercy, sympathy, pity, kindness; greeting; to love, show kindness, mercy; to venerate; to greet

alu: Combined; to cooperate, act together; excess

'ama'ama: Mullet (*Mugil cephalus*), a very choice indigenous fish

'anae: Full-sized *'ama'ama* mullet fish

'āpapa: Stratum, flat, especially a coral flat

'aumākua (*pl*), 'aumakua (*s*): Family or personal gods, deified ancestors who might assume the shape of sharks, owls, hawks, *'iwa* (frigatebirds), octopus, eels, or other animals

ea: Sovereignty, rule, independence; life, air, breath, respiration; to rise, go up

ēwe: Sprout, rootlet; lineage, kin; birthplace; family trait; to sprout

ha'awina: Lesson, assignment, task, allowance, grants, or contribution

hala: Pandanus tree (*Pandanus tectorius*), leaves (*lauhala*) are used for weaving; to pass, elapse, as time; to pass away

hale: House, building, institution, lodge, hall

hana: Work, labor, job, employment, occupation

haole: White person, American, Englishman, Caucasian; formerly, any foreigner

hāpai: To carry, bear, lift, elevate, raise; pregnant; to conceive

hā'uke'uke: An edible variety of sea urchin (*Colobocentrotus atratus*)

he'e: Octopus (*Polypus* spp.), commonly known as squid

hiki: Can, may; to be able; ability; possible

hoa ʻāina: Tenant, caretaker, as on a *kuleana*; *Lit.*, friend of the land

holoholo: To go for a walk, ride, or sail; to go out for pleasure

honu: General name for turtle and tortoise, as *Chelonia mydas*

hoʻokupu: To cause growth, sprouting; to sprout; tribute, tax, ceremonial gift-giving to a chief as a sign of honor and respect

hoʻomalu: To bring under the care and protection of, to protect

hoʻoponopono: To correct, amend; family conferences in which relationships were set right through a prayer, discussion, confession, repentance, and mutual restitution and forgiveness

hui: Club, association, organization, partnership, team; to form a society or group; to meet, congregate

huki: To pull or tug, as on a rope

hukilau: A seine; to fish with the seine; *Lit.*, pull ropes (*lau*)

iʻa: Fish or any marine animal, as eel, oyster, crab, whale

ʻili: Land section, next in importance to *ahupuaʻa* and usually a subdivision of an *ahupuaʻa*

imu: Underground oven

imu kai: Rock and coral fish trap; also *ahu, umu*

ʻiwa: Frigatebird or man-of-war bird (*Fregata minor palmerstoni*)

iwi: Bone; the bones of the dead, considered the most cherished possession

kāhea: To call, cry out, invoke, greet, name

kahu: Honored attendant, guardian, keeper, administrator, caretaker

kaiāulu: Community, neighborhood, village

kākoʻo: To uphold, support, favor, assist, prop up

kala: Surgeonfish, unicorn fish; Teuthidae: *Naso hexacanthus, N. unicornis, N. brevirostris*

kalo: Taro (*Colocasia esculenta*); Hawaiian staple crop cultivated for food and considered to be an ancestor; there are more than three hundred Hawaiian varieties; all parts of the plant are eaten, its starchy root as poi, *kulolo* (a dessert) or *ʻai paʻa* (steamed corm), and its leaves as *lūʻau*

kamaʻāina: Native-born, one born in a place; *Lit.*, land child

kamani: A large tree (*Calophyllum inophyllum*)

kanaka: Human being, person, mankind, population; subject, as of a chief; attendant or retainer in a family (often a term of affection or pride)

kapa: Tapa, as made from *wauke* or *māmaki* bark; formerly clothes of any kind or bedclothes

kapu: Taboo, prohibition; sacredness; forbidden; sacred, holy; no trespassing, keep out

kāpulu: Careless, slovenly, unclean, gross, untidy

keiki: Child, offspring, descendant, progeny

kiaʻi: Guard, watchman, caretaker; to watch, guard

kilo: To watch closely, examine, look around, observe

kīpuka: Variation or change of form, as a calm place in a high sea, deep place in a shoal, opening in a forest or in cloud formations, and especially a clear place or oasis within a lava bed where there is vegetation

 koʻa: Coral, coral head; fishing grounds, usually identified by lining up with marks on shore; fishing shrine

koaʻe: The tropic or boatswain bird, particularly the white-tailed tropicbird (*Phaethon lepturus dorotheae*), which inhabits cliffs

kōkua: Help, aid, assistance, assistant; to help, support

konohiki: Headman of an *ahupuaʻa* land division under the chief; land or fishing rights under control of the *konohiki*; such rights are sometimes called *konohiki* rights; *Lit.*, invites ability

kulāiwi: Native land, homeland; native

kūleʻa: Successful, competent; happily

kuleana: Right, privilege, concern, responsibility, title, jurisdiction, authority, ownership; reason, cause, function, justification; small piece of property, as within an *ahupuaʻa*

kumu: Teacher, tutor

kupa: Citizen, native; well-acquainted

kupa ʻāina: The native families of a place; *Lit.*, people familiar with a place

kūpuna (*pl*), kupuna (*s*): Grandparent, ancestor, relative or close friend of the grandparent's generation, grand-aunt, grand-uncle

kūʻula: Any stone god used to attract fish, whether tiny or enormous, carved or natural, named for the god of fishermen

kuʻuna: Place where a net is set in the sea; to let down a fish net

lāʻī: *Ti* leaf (contraction of *lau kī*).

lako: Supply, provisions, gear, fixtures, plenty; wealth; well-supplied, well-furnished, well-equipped; rich, prosperous

lānai: Porch, veranda, balcony, temporary roofed construction with open sides near a house.

lani: Sky, heaven; heavenly, spiritual

lau: Leaf, frond, leaflet, greens, *lauhala*: pandanus (family Pandanaceae, genus *Pandanus*) leaf

lawa: Enough, sufficient, ample; to have enough, be satisfied

lawaiʻa: Fisherman; fishing technique; to fish, to catch fish

lei: Garland, wreath; necklace of flowers, leaves, shells, or feathers, given as a symbol of affection. There are multiple techniques for making lei from strung to sewn, braided to wrapped. A beloved child or sweetheart; also to leap, fling, toss, or rise as a cloud

limu: A general name for all kinds of plants living under water, both fresh and salt; also algae growing in any damp place such as on the ground or on rocks

limu kala: Common, long, brown seaweeds (*Sargassum echinocarpum*) with stiff, toothed leaves; rarely eaten raw because of toughness; used in ceremonies to drive away sickness and obtain forgiveness limu kohu: A soft, succulent, small seaweed (*Asparagopsis taxiformis*), with branching and at times furry tops that are tan, pink, or dark red; one of the best-liked edible seaweeds

limu līpoa: Bladelike, branched, brown seaweeds (*Dictyopteris plagiogramma* and *D. australis*) with unique aroma and flavor

loʻi: Irrigated terrace, especially for taro

loko iʻa: Fishpond

lomi: To rub, press, squeeze, crush, knead, massage

lūʻau: Hawaiian feast, named for new taro leaves and the delicious food made from them

mahalo: Thanks, gratitude; to thank

māhele: Portion, division, section; land division of 1848 (the Great Māhele); to divide, apportion

maiau: Neat and careful in work; skillful, ingenious, expert; correct, careful, as in speech

makaʻāinana: People in general; citizen, commoner, subject; *Lit.*, people that attend the land

makaʻala: Alert, vigilant, watchful, wide awake

mākāhā: Sluice gate, as of a fishpond; entrance to or egress from an enclosure

makahiki/Makahiki: Year, age; annual; celebration and observance of the return of the rainy season; high surf and storms associated with the god Lonoikamakahiki, stretching roughly from October/November to January/February, marked with ceremony, games, and gatherings around the islands; a time of gratitude, harvest, cultivation of skills, and reflection.

makai: Ocean, toward the ocean

makua: Parent, any relative of the parents' generation, as uncle, aunt, cousin; main stalk of a plant; adult; full-grown, mature

māla ʻai: Garden, plantation, patch, cultivated field

mālama: To take care of, tend, attend, preserve, protect, maintain; to keep or observe, as a taboo

mana: Supernatural or divine power, authority, spiritual, privilege; to have power, authority

manini: Common reef surgeonfish (*Acanthurus triostegus*), also called convict tang

manō: Shark (general name)

māʻona: Satisfied after eating, full; to have eaten, to eat one's fill

mauka: Inland, toward the mountain

mele: Song, anthem, or chant of any kind; poem, poetry; to sing, chant

minamina: To regret, be sorry, deplore; to grieve for something that is lost

moena: Mat; couch, bed; resting place, hand woven mats made from *lauhala*

moi: Threadfish (*Polydactylus sexfilis*)

moku: District, island, islet, section

mōlī: Laysan albatross (*Phoebastria immutabilis*)

moʻo: Lizard, reptile of any kind, serpent; water spirit; succession, series, especially a genealogical line moʻokūʻauhau: Genealogy, genealogical story, family ancestry or lineage of teachers

moʻolelo: Story, tale, myth, history, tradition, literature, legend

moʻopuna: Grandchild; descendant; relatives two generations later, whether blood or adoptedmuʻumuʻu: A woman's underslip or chemise; a loose gown

naupaka: Native species of shrubs (*Scaevola*) found in mountains and near coasts, conspicuous for their white or light-colored half-flowers

nenue: Chub fish, also known as rudder or pilot fish (*Kyphosus bigibbus, K. vaigiensis*)

niuhi: Man-eating shark

ʻōahi: Ceremony performed in Hāʻena for special occasions such as the visit of an *aliʻi* or other special guests

ʻoama: Young of the *weke*, goatfish

ʻohana: Family, relative, kin group; related

ʻōʻio: Ladyfish, bonefish (*Albula vulpes*)

ʻōiwi: Native, native son; *Lit.*, one whose ancestors bones are in the land

ʻōlelo: Language, speech, word; to speak, talk, converse

oli: Chant; many different styles of chant for different occasions and purposes

olonā: A native shrub (*Touchardia latifolia*); formerly the bark was valued highly as the source of a strong, durable fiber for fishing nets, and for nets (*kōkō*) to carry containers

ʻono: Delicious, tasty, savory; to relish, crave

ʻōpala: Trash, rubbish, refuse, litter

ʻōpelu: Mackerel scad (*Decapterus pinnulatus* and *D. maruadsi*)

ʻopihi: Limpets; Hawaiians recognize three kinds: *kōʻele* (*Cellana talcosa*, the largest), *ʻālinalina* (*C. sandwicensis*), and *makaiauli* (*C. exarata*)

Organic Act: An act that established Hawaiʻi as a territory of the United States on April 30, 1900; included in sections 95 and 96 was language to repeal laws conferring exclusive *konohiki* fishing rights subject to vested rights and proceedings for opening fisheries to all citizens

pae ʻāina: Group of islands, archipelago

palena: Boundary, limit, border

palu: A delicacy made of fish flavored with innards and other parts, mixed with *kukui* relish, garlic, chili peppers; also bait used to chum for fish

Papahānaumoku: Earth mother, sometimes called Papa, mother of islands

pau: Finished, ended, through, terminated, completed, over, all done

pele: lava flow, volcano, eruption; volcanic (named for the volcano goddess, Pele)

piko: Navel, umbilical cord; center, source

pili: A grass (*Heteropogon contortus*), formerly used for thatching houses in Hawaiʻi; to cling or be close, *pilina* (relationship)

pōhaku: Rock, stone

pōhuehue: The beach morning-glory (*Ipomoea pes-caprae* subsp. *brasiliensis*), a strong vine found on sandy beaches in the tropics

poi: The Hawaiian staff of life, made from cooked taro corms, or sometimes breadfruit, pounded and thinned with water poke: To slice, cut crosswise into pieces, as fish or wood

pono: Goodness, uprightness, correct or proper procedure; moral, fitting, proper, righteous

puakenikeni: A shrub or small tree (*Fagraea berteriana*), grown ornamentally for foliage, flowers, fruit; *Lit.*, ten-cent flower

pueo: Hawaiian short-eared owl (*Asio flammeus sandwichensis*), regarded as 'aumākua

puhi: Eels, in general (Muraenidae, Ophichthidae, Congridae)

puka: Hole; door, entrance, gate, slit, vent, opening

pule: Prayer, blessing

pūʻolo: Bundle, container, parcel. *Ti* leaf bundle used to wrap a lei or other offering

ti or kī: A woody plant (*Cordyline terminalis*); Hawaiians use the leaves for many purposes including for house thatch, food wrappers, medicine and hula skirts

tūtū: Grandma; grandpa

ʻulu: The breadfruit (*Artocarpus altilis*), a tree grown for its edible fruits

ulua: Certain species of Carangidae (crevalle, jack, or pompano), the most common is the giant trevally (*Caranx ignobilis*), an important game fish and food item

ʻupena: Fishing net, net, web

wahine: Woman

waiwai: Goods, property, assets, value, worth, wealth, abundant water

weke: Certain species of the Mullidae, surmullets or goatfish

wili: To wind, twist, writhe, crank, turn; wrapped style of lei making

Works Cited

Act of Sept. 1, 1950, Chapter 833, 64 Stat. 572. Legislature of the Territory of Hawai'i.

Act of Aug. 23, 1954, Chapter 824, 68 Stat. 764. Legislature of the Territory of Hawai'i.

Act 241 Establishing the Hā'ena CBSFA. 2006. State of Hawai'i, Honolulu. http://dlnr .hawaii.gov/dar/files/2016/03/Haena_CBSFA_Draft_Mgmt_Plan_3.1.16.pdf.

Agrawal, A., and C. C. Gibson. 1999. Enchantment and Disenchantment: The Role of Community in Natural Resource Conservation. *World Development* 27 (4): 629–649.

Akutagawa, M. 2016. Appendix G: Evaluation of Proposed Hawai'i Noncommercial Marine Fishing Registry, Permit, and License Design Scenarios and Policy Recommendations for Resolving Potential Conflicts with Native Hawaiian Rights. In *Feasibility of a Non-Commercial Fishing Registry, Permit, or License System in Hawai'i—Study Group Final Report*. Prepared for Conservation International-Hawai'i. http://dlnr.hawaii.gov/dar/announcements/feasibility-of-a-non-commercial -marine-fishing-registry-permit-or-license-system-in-hawaii/.

———. Forthcoming 2018. An Island Negotiating a Pathway for Responsible Tourism. In *Detours: A Decolonial Guidebook to Hawai'i*, edited by Hokulani Aikau and Vernadette Gonzalez.

Akutagawa, M., E. Cole, T. P. Diaz, T. D. Gupta, C. Gupta, A. Fa'anunu, S. Kamaka'ala, and M. Tauali'i. 2016. *Health Impact Assessment of the Proposed Mo'omomi Community-Based Subsistence Fishing Area—Island of Moloka'i, Hawai'i*. Prepared for the Kohala Center. http://kohalacenter.org/docs/reports/Moomomi_HIA _FullReport_Web_Final.pdf.

Akutagawa, M., H. Williams, and S. Kamaka'ala. 2016. *Traditional and Customary Practices Report for Mana'e, Moloka'i Traditional Subsistence Uses, Mālama Practices and Recommendations, and Native Hawaiian Rights Protections of Kama'āina Families of Mana'e Moku, East Moloka'i, Hawai'i*. Honolulu: University of Hawai'i at Mānoa William S. Richardson School of Law.

Alayvilla, A. 2016. Number of Drownings Off Kauai Nearly Doubles Last Year's Figure. *Garden Island*, November 6. http://thegardenisland.com/news/local/number-of -drownings-off-kauai-nearly-doubles-last-year-s/article_9d188664-0735-587f -a180-6c629e6bbb62.html.

Alfred, T., and J. Corntassel. 2005. Being Indigenous: Resurgences against Contemporary Colonialism. *Government and Opposition* 40 (4): 597–614.

Anderson, M. K. 2013. *Tending the Wild: Native American Knowledge and the Management of California's Natural Resources*. Reprint ed. Berkeley: University of California Press.

Andrade, C. 2008. *Hāʻena: Through the Eyes of the Ancestors*. Honolulu: University of Hawaiʻi Press.

Armitage, D., F. Berkes, A. Dale, E. Kocho-Schellenberg, and E. Patton. 2011. Co-management and the Co-production of Knowledge: Learning to Adapt in Canada's Arctic. *Global Environmental Change* 21 (3): 995–1004.

Ayers, A. L., and J. N. Kittinger. 2014. Emergence of Co-management Governance for Hawaiʻi Coral Reef Fisheries. *Global Environmental Change* 28:251–262.

Azambuja, L. 2011. DLNR Boosting Fisheries Enforcement. *Garden Island*, May 26. http://thegardenisland.com/news/local/govt-and-politics/article_d27d1e1a-883a-11e0-9883-001cc4c03286.html.

Bakunin, M. 1882. What Is Authority? http://www.panarchy.org/bakunin/authority.1871.html.

Barrera, W. 1984. Wainiha Valley, Kauai: Archaeological Studies. Prepared for Orion Engineering. Honolulu: Chiniago Enterprises.

Bay, J. H., and J. van Schaick. 1994. Analysis of the 999-Year Homestead Lease Program: Current Problems and Possible Solutions. Prepared for the Seventeenth Legislature of the State of Hawaiʻi and the Office of Hawaiian Affairs.

Beamer, K. 2014. *No Mākou Ka Mana: Liberating the Nation*. Honolulu: Kamehameha Schools Press.

Berkes, F. 2004. Rethinking Community-Based Conservation. *Conservation Biology* 18 (3): 621–630.

———. 2009. Evolution of Co-management: Role of Knowledge Generation, Bridging Organizations and Social Learning. *Journal of Environmental Management* 90 (5): 1692–1702.

———. 2015. *Coasts for People: Interdisciplinary Approaches to Coastal and Marine Resource Management*. New York: Routledge.

Berkes, F., J. Colding, and C. Folke. 2000. Rediscovery of Traditional Ecological Knowledge as Adaptive Management. *Ecological Applications* 10 (5): 1251–1262.

Blaich, B., and B. Robeson. 1986. Hanalei Yesterday. Historic Hanalei Project, unpublished flyer.

Blaich, M. 2003. Mai Uka a i Kai: From the Mountains to the Sea: ʻĀina Based Education in the Ahupuaʻa of Waipā. Unpublished master's thesis, University of Hawaiʻi at Mānoa, Honolulu.

Blay, C. T. 2011. Drowning Deaths in the Nearshore Marine Waters of Kauai, Hawaii 1970–2009. *International Journal of Aquatic Research and Education* 5 (3): 284–324.

Brownell, R. L. Jr., K. Ralls, S. Baumann-Pickering, and M. M. Poole. 2009. Behavior of Melon-Headed Whales, *Peponocephala electra*, Near Oceanic Islands. *Marine Mammal Science* 25 (3): 639–658.

Buley, B. 2017. Whales Beach at Kalapaki. *Garden Island*, October 14. http

://thegardenisland.com/news/local/whales-beach-at-kalapaki/article_a9231aa8
-bb7e-58dc-a4d5-a06b1dcdc78c.html.

Cadiz, E., N. Peralto, and A. Kagawa. 2015. An Overview of Cultural Resources Relating to Anini, Haleleʻa, Kauaʻi: Including the ʻIli of Anini, Hanalei, and Portions of the Ahupuaʻa of Kalihikai and Kalihiwai. Unpublished manuscript.

Callicott, J. B. 1994. *Earth's Insights: A Multicultural Survey of Ecological Ethics from the Mediterranean Basin to the Australian Outback*. Berkeley: University of California Press.

Carpenter, J. 2017. Kauai Population Tops 72,000. *Garden Island*, March 28. http ://thegardenisland.com/news/local/kauai-population-tops/article_776a652f-0150-5a70-8726-cfacf978e1c6.html.

Central Pacific Hurricane Center. n.d. The 1992 Central Pacific Tropical Cyclone Seas. National Oceanic and Atmospheric Administration. http://www.prh.noaa.gov /cphc/summaries/1992.php.

Chun, G. 2009. Remembering Our Future: Moving beyond Sustainability. Speech presented at the Kohala Center. http://kohalacenter.org/liem-hawaii/presentations/ chun.

Cinner, J. E., and S. Aswani. 2007. Integrating Customary Management into Marine Conservation. *Biological Conservation* 140 (3): 201–216.

Civil Beat Editorial Board. 2016. Hawaiʻi Leaders Must Do More to Prevent Tourist Deaths in 2016. *Civil Beat*, January 19. http://www.civilbeat.org/2016/01 /hawaii-leaders-must-do-more-to-prevent-tourist-deaths-in-2016/.

Cooper, G., and G. Daws. 1990. *Land and Power in Hawaii: The Democratic Years*. Honolulu: University of Hawaiʻi Press.

Cordy, D. 2016. Where Are the Crown Lands of Hawaiʻi? Remapping ʻĀina. Unpublished master's thesis, University of Hawaiʻi at Mānoa.

Corntassel, J. 2012. Re-envisioning Resurgence: Indigenous Pathways to Decolonization and Sustainable Self-Determination. *Decolonization: Indigeneity, Education and Society* 1 (1): 86–101.

County of Kauaʻi. 2016. 2016 Tax Rates. www.kauai.gov/Government/ Departments-Agencies/Finance/Real-Property/Tax-Rates.

County of Kauaʻi. 2017. Kauaʻi Kākou, Kauaʻi County General Plan, Planning Commission Draft. Prepared for the Kauaʻi County Council. https://www.dropbox .com/s/zmc323or4pm8kvq/Kauai%20GP%20Body%20Combined%20170720%20 reduced%20size.pdf?dl=0.

Cox, M., G. Arnold, and S. Villamayor Tomás. 2010. A Review of Design Principles for Community-Based Natural Resource Management. *Ecology and Society* 15 (4): 38. http://www.ecologyandsociety.org/vol15/iss4/art38/.

DeLoria, Vine. 2006. *The World We Used to Live In: Remembering the Powers of the Medicine Men*. Golden, CO: Fulcrum.

Department of Business Economic Development and Tourism (DBEDT). 2016. Residential Home Sales in Hawaiʻi, Trends and Characteristics: 2008–2015.

http://files.hawaii.gov/dbedt/economic/data_reports/homesale/Residential_Home
_Sales_in_Hawaii_May2016.pdf.

———. 2017. *The State of Hawai'i Data Book 2016.* http://files.hawaii.gov/dbedt
/economic/databook/db2016/db2016.pdf.

Diver, Sibyl. 2012. Columbia River Tribal Fisheries: Life History Stages of a Co-
Management Institution. In *Keystone Nations: Indigenous Peoples and Salmon across
the North Pacific,* edited by Benedict J. Colombi and James Brooks, p. 28. Santa Fe,
NM: School for Advanced Research Press.

Doughty, A. 2014. *The Ultimate Kauai Guidebook: Kauai Revealed.* Lihue, HI: Wizard
Publications.

Drewes, P. 2017. What Would Happen If Iniki Hit Oahu? KITV News, May 31.
http://www.kitv.com/story/35552032/what-would-happen-if-iniki-hit-oahu.

Dye, K. P., and T. S. Dye. 2010. *An Archaeological Survey for Animal Control Fencing in
the Wainiha Preserve, Wainiha Valley, Kauai.* Honolulu: T. S. Dye and Colleagues,
Archaeologists.

Earle, T. K. 1978. Economic and Social Organization of a Complex Chiefdom:
The Halele'a District, Haua'i, Hawai'i. In *Anthropological Papers* 63. Ann Arbor:
University of Michigan Museum of Anthropology.

Elbert, S. H. 1959. *Selections from Fornander's Hawaiian Antiquities and Folk-Lore.*
Honolulu: University of Hawai'i Press.

Emerson, N. B. 1909. Unwritten Literature of Hawaii, Bureau of American
Ethnology. *Bulletin 38.*

Fitzsimmons, J. M., J. E. Parham, L. K. Benson, M. G. McRae, and R. T. Nishimoto. 2005.
Biological Assessment of Kahana Stream, Island of O'ahu, Hawai'i: An Application of
PABITRA Survey Methods. *Pacific Science* 59 (2): 273–281.

Friedlander, A. M., and K. S. Rodgers. 2008. Coral Reef Fishes and Fisheries of South
Moloka'i. In *The Coral Reef of South Moloka`i, Hawai'i; Portrait of a Sediment-
Threatened Fringing Reef,* edited by M. E. Field, S. A. Cochran, J. B. Logan, and C. D.
Storlazzi, pp. 59–66. US Geological Survey Scientific Investigations Report 2007-5101.

Friedlander, A. M., J. M. Shackeroff, and J. N. Kittinger. 2013. Customary Marine
Resource Knowledge and Use in Contemporary Hawai'i. *Pacific Science* 67 (3):
441–460.

Fu, K., B. Kinney, and M. Vaughan. 2015. Aloha Anini e Waiho nei: Stories of Wanini,
Kalihikai, and Hanapai, Kaua'i. Unpublished Report.

Garovoy, J. B. 2005. Ua Koe Ke Kuleana O Na Kanaka (Reserving the Rights of Native
Tenants): Integrating Kuleana Rights and Land Trust Priorities in Hawai'i. *Harvard
Environmental Law Review* 29:523.

Gomes, A. 2017. Facebook Co-founder Wins Praise for Decision to End Suits. *Honolulu
Star Advertiser,* January 28. http://www.staradvertiser.com/2017/01/28/business/
facebook-co-founder-wins-praise-for-decision-to-end-suits/.

Gonschor, L., and K. Beamer. 2014. Toward an Inventory of *Ahupua'a* in the Hawaiian
Kingdom: A Survey of Nineteenth- and Early Twentieth-Century Cartographic
and Archival Records of the Island of Hawai'i. *Hawaiian Journal of History*
48:53–87.

Goodyear-Kaʻōpua, N. 2011. Kuleana Lahui: Collective Responsibility for Hawaiian Nationhood in Activists' Praxis. *Affinities: A Journal of Radical Theory, Culture, and Action* 5 (1): 130–163.

———. 2013. *The Seeds We Planted: Portraits of a Native Hawaiian Charter School.* Minneapolis: University of Minnesota Press.

Hāʻena Community. 2000. Hāʻena Community Study of Residences. Unpublished data.

Hall, D. N. 2017. *Life of the Land: Articulations of a Native Writer.* Honolulu: ʻAi Pohaku Press.

Handy, E. S. C. 1985. *Polynesian Religion.* Millwood, NY: Periodicals Service Co.

Hawaiʻi Administrative Rules (HAR). 2014. Title 13 Department of Land and Natural Resources, Subtitle 4 Fisheries, Part II Marine Fisheries Management Areas, Chapter 60.8 Hāʻena Community-Based Subsistence Fishing Area, Kauaʻi.

Hawaiʻi Revised Statutes (HRS). 1959. Section 5-9. State Motto. Title 1. General Provisions, Chapter 5. Emblem and Symbols. Honolulu: State of Hawaiʻi.

———. 2015a. Section 187A-23. Konohiki Rights. Title 12. Conservation and Resources, Chapter 187A. Aquatic Resources. Honolulu: State of Hawaiʻi.

———. 2015b. Section 188-22.9. Hāʻena Community-Based Subsistence Fishing Area; Restrictions; Regulations. Title 12. Conservation and Resources, Chapter 188. Fishing Rights and Regulations. Honolulu: State of Hawaiʻi.

Heat-Moon, L. 1999. *PrairyErth: An Epic History of the Tallgrass Prairie Country.* Rev. ed. New York: Houghton Mifflin/First Mariner.

Higuchi, J. 2008. Propagating Cultural *Kipuka*: The Obstacles and Opportunities of Establishing a Community-Based Subsistence Fishing Area. *University of Hawaii Law Review* 31:193.

Holmes, M. C. C. 2010. Law for Country: The Structure and Application of Warlpiri Ecological Knowledge. PhD thesis, School of English, Media Studies and Art History, University of Queensland. http://espace.library.uq.edu.au/view/UQ:239731.

Hoʻoulumahiehie, H., and M. P. Nogelmeier. 2006. *Ka moʻolelo o Hiʻiakaikapoliopele: ka wahine i ka hikina a ka lā, ka uʻi palekoki uila o Halemaʻumaʻu.* Honolulu: Awaiaulu. http://natlib.govt.nz/records/21433650.

Johannes, R. E. 2002. The Renaissance of Community-Based Marine Resource Management in Oceania. *Annual Review of Ecology and Systematics* 33 (1): 317–340. http://doi.org/10.1146/annurev.ecolsys.33.010802.150524.

Jokiel, P. L., K. S. Rodgers, W. J. Walsh, D. A. Polhemus, and T. A. Wilhelm. 2011. Marine Resource Management in the Hawaiian Archipelago: The Traditional Hawaiian System in Relation to the Western Approach. *Journal of Marine Biology.* doi:10.1155/2011/151682.

Kahaulelio, A. D., and P. Nogelmeier. 2005. *Ka Oihana Lawaia: Hawaiian Fishing Traditions.* Honolulu: Bishop Museum Press.

Kameʻeleihiwa, L. 1992. *Native Land and Foreign Desires: How Shall We Live in Harmony?* Honolulu: Bishop Museum Press.

Kanahele, P. K., R. S. Bergman, K. H. Muhammad, and S. Syamsuddin. 2016. Connections: Spirituality and Conservation. High-level discussion panel conducted at the International Union of the Conservation of Nature World Conservation Congress, Honolulu, September 5.

Kaomea, J. 2014. Reading Erasures and Making the Familiar Strange: Defamiliarizing Methods for Research in Formerly Colonized and Historically Oppressed Communities. *Educational Researcher* 32 (2): 14–25.

Kawelo, H. 2015. Kilo. ʻAimalama: Pacific Peopleʻs Lunar Conference on Climate Change. Keoni Auditorium, East-West Center, UH-Mānoa, September 25.

Kazmirzack, R. 2015. Revealing More about Hawaiʻi Revealed. *Garden Island*, June 14. http://thegardenisland.com/news/local/revealing-more-about-hawaii-revealed/article_4b0d5dc8-780d-586f-a0e6-e200cc152690.html.

Keale, W. 2009. Pule Hoʻola (Na ʻAumākua). On *Kawelona: Ride The Sun*. Kaneohe, Hawaiʻi: Rhythm and Roots Records. https://store.cdbaby.com/cd/keale2.

Kimmerer, R. W. 2013. *Braiding Sweetgrass: Indigenous Wisdom, Scientific Knowledge and the Teachings of Plants*. Minneapolis, MN: Milkweed Editions.

Kirch, P. V., and J. L. Rallu. 2007. *The Growth and Collapse of Pacific Island Societies: Archaeological and Demographic Perspectives*. Honolulu: University of Hawaiʻi Press.

Kittinger, J. N., A. L. Ayers, and E. E. Prahler. 2012. Policy Briefing: Co-Management of Coastal Fisheries in Hawaii: Overview and Prospects for Implementation. Monterey, CA: Center for Ocean Solutions, Stanford University. http://papers.ssrn.com/sol3/papers.cfm?abstract_id¼2590207.

Koethe, F. S. 2017. Community Perceptions and Priorities for Sustainability in Anini, Kauaʻi. Masterʻs thesis, University of Hawaiʻi at Mānoa.

Koethe, F., J. Muratsuchi, W. Sowa, T. Taylor, and R. Vave. 2015. An Assessment of the Environmental and Natural Resource History of a Coastal Hawaiʻi Community: A Case Study of Anini, Kauaʻi. Unpublished manuscript.

Kosaki, R. H. 1954. *Konohiki Fishing Rights, Hawaiʻi Legislature. No. 1 (Request No. 3642)*. Honolulu: University of Hawaiʻi, Legislative Reference Bureau.

Kuykendall, R. S. 1938. *The Hawaiian Kingdom*. Vol. 1. Honolulu: University of Hawaiʻi Press.

Lukacs, H. A., and N. M. Ardoin. 2014. The Relationship of Place Re-Making and Watershed Group Participation in Appalachia. *Society and Natural Resources* 27 (1): 55–69.

Liliʻuokalani, Queen. 1978. *The Kumulipo: An Hawaiian Creation Myth*. Kentfield, CA: Pueo Press.

Magnusson, W., and K. Shaw. 2002. *A Political Space: Reading the Global through Clayoquot Sound*. Ontario: McGill-Queenʻs University Press.

Maly, K., and O. Maly. 2003. "Hana Ka Lima, Ai Ka Waha": A Collection of Historical Accounts and Oral History Interviews with Kamaʻaina Residents and Fisher-People of Lands in the Haleleʻa-Napali Region. Prepared for the Nature Conservancy.

Mamaril, N. M., L. J. Cox, and M. Vaughan. Forthcoming. *Weaving Evaluation into the Waipā Ecosystem: Placing Evaluation in an Indigenous Place-Based Education Program*.

Manu, M., and Others. 2006. *Hawaiian Fishing Traditions: Revised Edition 2006*. Honolulu: Kalamaku Press.

McClenachan, L., and J. N. Kittinger. 2012. Multicentury Trends and the Sustainability of Coral Reef Fisheries in Hawaiʻi and Florida. *Fish and Fisheries* 14 (3): 239–255.

McCracken, D. 2017. Flights to Kauai to Increase by 42.6 Percent in 2018. *Garden Island*, August 27. http://thegardenisland.com/news/local/flights-to-kauai-to-increase -by-percent-in/article_c7c3fc4c-4b9f-5ace-9668-50d72be83920.html.

McGregor, D. P. 2006. Cultural Assessment for the Kamakou Preserve, Makaupaʻia and Kawela, Island of Molokaʻi. Prepared for The Nature Conservancy.

———. 2007. *Na Kuaʻaina: Living Hawaiian Culture*. Honolulu: University of Hawaiʻi Press.

McGregor, D. P., and M. K. Mackenzie. 2014. *Moʻolelo Ea o Hawaii. History of Native Hawaiian Governance*. Prepared for the Office of Hawaiian Affairs.

Menzies, N. K. 2007. *Our Forest, Your Ecosystem, Their Timber: Communities, Conservation, and the State in Community-Based Forest Management*. New York: Columbia University Press.

Montgomery, M. 2018. Lessons from Konohiki: Building Capacity for Community-Based Coastal Resource Management. Unpublished master's thesis, University of Hawaiʻi at Mānoa.

Nakuina, E. M. 1994. *Nanaue the Shark Man and Other Hawaiian Shark Stories*. D. Kawaharada, ed. Honolulu: Kalamakū Press.

Okri, B. 1996. *Birds of Heaven*. London: Phoenix.

Oliveira, K. N. 2014. *Ancestral Places: Understanding Kanaka Geographies*. 1st ed. Corvallis: Oregon State University Press.

Ostrom, E. 1990. *Governing the Commons: The Evolution of Institutions for Collective Action*. New York: Cambridge University Press.

———. 2005. *Understanding Institutional Diversity*. Princeton, NJ: Princeton University Press.

Ostrom, E., J. Burger, C. B. Field, R. B. Norgaard, and D. Policansky. 1999. Revisiting the Commons: Local Lessons, Global Challenges. *Science* 284 (5412): 278–282.

Pa, Chauncey. 2010. The Life and Times of Chauncey Wightman Pa. Unpublished manuscript.

Panelli, R., and G. Tipa. 2009. Beyond Foodscapes: Considering Geographies of Indigenous Well-Being. *Health and Place* 15 (2): 455–465.

Pascua, P. A., H. McMillen, T. Ticktin, M. Vaughan, and K. B. Winter. 2017. Beyond Services: A Process and Framework to Incorporate Cultural, Genealogical, Place-Based, and Indigenous Relationships in Ecosystem Service Assessments. *Ecosystem Services* 26 (B): 465–475.

Paulo, W. K., D. K. Seto, E. M. Enos, P. Burgess, and WCCAD. 1996. *From Then to Now: A Manual for Doing Things Hawaiian Style*. Waiʻanae, Hawaiʻi: Waiʻanae Coast Community Alternative Development Corp.

Peralto, L. N. 2014. ʻO Koholālele, He ʻĀina, He Kanaka, He I ʻa Nui Nona ka Lā: Remembering Knowledge of Place in Koholālele, Hāmākua, Hawaiʻi. In *I Ulu I Ka ʻAina: Land*. Honolulu: Hawaiʻinuiākea School of Hawaiian Knowledge.

———. Forthcoming 2018. *Kokolo mai ka mole uaua o 'Ī: The Resilience and Resurgence of Aloha 'Āina in Hāmākua Hikina, Hawai'i.* Dissertation, Department of Political Science, Indigenous Politics Program. University of Hawai'i at Mānoa.

Pinkerton, E., and R. Davis. 2015. Neoliberalism and the Politics of Enclosure in North American Small-Scale Fisheries. *Marine Policy* 61:303–312.

Poepoe, K. K., P. K. Bartram, and A. M. Friedlander. 2007. The Use of Traditional Knowledge in the Contemporary Management of a Hawaiian Community's Marine Resources. In *Fishers' Knowledge in Fisheries Science and Management*, edited by N. Haggan, B. Neis, and I. G. Baird. Oxford: Blackwell Science/Paris: UNESCO.

Preza, D. 2010. The Empirical Writes Back: Re-examining Hawaiian Dispossession Resulting from the Māhele of 1848. Unpublished doctoral dissertation, University of Hawai'i at Mānoa.

Pukui, M. K. 1983. *'Ōlelo No'eau: Hawaiian Proverbs and Poetical Sayings.* Vol. 71. Honolulu: Bishop Museum Press.

Pukui, M., E. Haertig, and C. Lee. 1979. *Nana I Ke Kumu (Look to the Source).* Honolulu: Hui Hanai.

Pukui, M. K., and S. H. Elbert. 1986. *Hawaiian Dictionary: Hawaiian-English, English-Hawaiian.* Honolulu: University of Hawai'i Press.

Putnam, R. D. 1995. Bowling Alone: America's Declining Social Capital. *Journal of Democracy* 6 (1): 65–78.

Raibmon, P. 2005. *Authentic Indians: Episodes of Encounter from the Late-Nineteenth-Century Northwest Coast.* Durham, NC: Duke University Press.

Ribot, J. C., and N. L. Peluso. 2003. A Theory of Access. *Rural Sociology* 68 (2): 153–181.

Richmond, L. 2014. *Anagyuk* (Partner): Personal Relationships and the Exploration of Sugpiaq Fishing Geographies in Old Harbor, AK. In *A Deeper Sense of Place: New Geographies of Indigenous-Academic Collaboration*, edited by J. Johnson and S. Larson. Corvallis: Oregon State University Press.

Schemmel, E. M., and A. M. Friedlander. 2017. Participatory Fishery Monitoring is Successful for Understanding the Reproductive Biology Needed for Local Fisheries Management. *Environmental Biology of Fishes* 100 (2): 171–185.

Schewe, R. L., D. R. Field, D. J. Frosch, G. Clendenning, and D. Jensen. 2012. *Condos in the Woods: The Growth of Seasonal and Retirement Homes in Northern Wisconsin.* Madison: University of Wisconsin Press.

Shimogawa, D. 2014. Kauai's Princeville Resort's $500M Redevelopment to Create 600 Jobs. *Pacific Business News*, November 20. http://www.bizjournals.com/pacific/news/2014/11/20/kauais-princeville-resorts-500m-redevelopment-to.html.

———. 2016. Facebook's Mark Zuckerberg to Build Home on Kauai Oceanfront Estate. *Pacific Business News*, June 30. http://www.bizjournals.com/pacific/news/2016/06/29/facebooks-mark-zuckerberg-to-build-home-on-kauai.html.

———. 2017. Sylvester Stallone's Former Hawai'i Estate Sold for $11 Million. *Pacific Business News*, March 27. http://www.bizjournals.com/pacific/news/2017/03/27/sylvester-stallones-former-hawaii-estate-sold-for.html.

Shomura, R. S. 1987. *Hawaii's Marine Fishery Resources: Yesterday (1900) and Today (1986)*. Southwest Fisheries Center, National Marine Fisheries Service, Honolulu Laboratory.

Simpson, L. 2011. *Dancing on Our Turtle's Back: Stories of Nishnaabeg Re-creation, Resurgence and a New Emergence*. Winnipeg, Canada: Arbeiter Ring.

Solis, S. C. K. 2013. Wainiha Hui Kūʻai ʻĀina, Ancestral Lands Forever: A *Moʻolelo* of Kānaka and *ʻĀina* Persistence. Master's thesis, University of Hawaiʻi at Mānoa.

Sproat, D. 2010. Where Justice Flows Like Water: The Moon Court's Role in Illuminating Hawaiʻi Water Law. *University of Hawaiʻi Law Review* 33:537.

———. 2016. An Indigenous People's Right to Environmental Self-Determination: Native Hawaiians and the Struggle against Climate Change Devastation. *Stanford Environmental Law Journal* 35:157.

State of Hawaiʻi. 2017. About DLNR. http://dlnr.hawaii.gov/about-dlnr/.

Stauffer, R. H. 2004. *Kahana: How the Land Was Lost*. Honolulu: University of Hawaiʻi Press.

Steele, J. 2016. Episode 40: Literacy and Hawaiian-Language Newspapers with Kauʻi Sai-Dudoit. http://hawaiipublicradio.org/post/episode-40-literacy-and-hawaiian-language-newspapers-kau-i-sai-dudoit.

Stepath, C. M. 2006. *Kēʻē Lagoon and Reef Flat Users Baseline Study*. Unpublished report.

Tipa, G., and R. Panelli. 2009. Beyond "Someone Else's Agenda": An Example of Indigenous/Academic Research Collaboration. *New Zealand Geographer* 65 (2): 95–106.

Tipa, G., and R. Welch. 2006. Comanagement of Natural Resources Issues of Definition from an Indigenous Community Perspective. *Journal of Applied Behavioral Science* 42 (3): 373–391.

Tong, N. 2014. He ʻĀina Wai: Remembering Water Narratives of Waiʻanae Kai. Unpublished master's thesis, University of Hawaiʻi at Mānoa, Honolulu.

Turner, N. J. 2008. *The Earth's Blanket: Traditional Teachings for Sustainable Living*. Madeira Park, BC: D&M Publishers.

Turner, N. J., M. B. Ignace, and R. Ignace. 2000. Traditional Ecological Knowledge and Wisdom of Aboriginal Peoples in British Columbia. *Ecological applications* 10 (5): 1275–1287.

Turner, N., and P. Spalding. 2013. "We Might Go Back to This": Drawing on the Past to Meet the Future in Northwestern North American Indigenous Communities. *Ecology and Society* 18 (4): 29.

United States. 1911. Organic Act of the Territory of Hawaii (As Amended). Annotated. Honolulu: Bulletin Publishing. http://darrow.law.umn.edu/documents/Organic_Act.pdf.

US Census Bureau. 2016. American Fact Finder, Estimates of the Components of Resident Population Change: April 1, 2010 to July 1, 2016, 2016 Population Estimates. http://files.hawaii.gov/dbedt/census/popestimate/2016-county-population-hawaii/PEP_2016_PEPTCOMP.pdf.

Vaughan, M. B. 2014. ʻĀina (Land), That Which Feeds: Researching Community Based Natural Resource Management at Home. *Journal of Research Practice* 10 (2): 19.

Vaughan, M. B., and P. M. Vitousek. 2013. Mahele: Sustaining Communities through Small-Scale Inshore Fishery Catch and Sharing Networks 1. *Pacific Science* 67 (3): 329–344.

Vaughan, M. B., and N. M. Ardoin. 2014. The Implications of Differing Tourist/Resident Perceptions for Community-Based Resource Management: A Hawaiian Coastal Resource Area Study. *Journal of Sustainable Tourism* 22 (1): 50–68.

Vaughan, M. B., and M. R. Caldwell. 2015. Hana Paʻa: Challenges and Lessons for Early Phases of Co-management. *Marine Policy* 62:51–62.

Vaughan, M. B., and A. L. Ayers. 2016. Customary Access: Sustaining Local Control of Fishing and Food on Kauaʻiʻs North Shore. *Food, Culture and Society* 19 (3): 517–538.

Vaughan, M. B., B. Thompson, and A. L. Ayers. 2016. Pāwehe Ke Kai a ʻo Hāʻena: Creating State Law Based on Customary Indigenous Norms of Coastal Management. *Society and Natural Resources* 30 (1): 1–16.

Veracini, L. 2011. Introducing Settler Colonial Studies. *Settler Colonial Studies* 1 (1): 1–12.

Waipā Foundation. n.d. Community. www.waipafoundation.org.

Wichman, F. B. 1998. *Kauaʻi: Ancient Place-Names and Their Stories*. Honolulu: University of Hawaiʻi Press.

Wilson, S. 2009. *Research Is Ceremony: Indigenous Research Methods*. Halifax, Nova Scotia: Fernwood.

Wolfe, P. 2006. Settler Colonialism and the Elimination of the Native. *Journal of Genocide Research* 8 (4): 387–409.

Wright, E. K. 2014. Kuleana Acts: Identity in Action. In *Indigenous Leadership in Higher Education*, edited by Robin Minthorn and Alicia Fedelina Chavez, pp. 127–135. New York: Routledge.

Zuckerberg, M. 2017. Zuckerberg Vows to Be Good Steward on Kauai. *Garden Island*, January 28. http://thegardenisland.com/news/opinion/guest/zuckerberg-vows-to-be-good-steward-on-kauai/article_47837f8d-8371-5a4f-807a-624643e5ef5a.html.

Mahalo Piha (Acknowledgments)

Ola i ke ahe lau makani.
There is life in a gentle breath of wind.
(Said when a warm day is relieved by a breeze.)
—Pukui 1983, p. 271

Bringing this book to life over the past twenty years has taken a community. First, *mahalo* to the *'ohana* of Ko'olau and Halele'a and all who have turned hands down to *mālama* these verdant lands, mountain to sea, past to present. *Mahalo* to those who have shared their families' stories and teachings with such love and generosity. This is your book, may it be enjoyed with your children and grandchildren. Fond memories and gratitude to the many places (Mānoa, Wai'anae, Miloli'i, Kaua'i o Kamawaelualani) and *kūpuna* who have been my teachers. I have been indelibly shaped by my time with my Tūtū Amelia Ana Ka'ōpua Bailey, her sisters Aunty Nancy Ahana and Aunty Marcella Brede, Uncle Eddie Ka'anana, Uncle Walter Paulo, Aunty Loke Pereira, Uncle Valentine Ako, Aunty Annabelle Kam, Uncle Henry Chang-Wo, Aunty Puanani Burgess, and many more.

Mahalo to Hā'ena's Hui Maka'āinana o Makana, Limahuli Garden 'ohana, Waipā Foundation, Mālama Kalihiwai, Nā Kia'i o Nihokū, and the many longtime and young adult community leaders for the unique perspectives you have each added to this work. Gratitude to all who took time to come to research-sharing gatherings and to bring food. *Mahalo* to Kua 'Āina Ulu 'Auamo (KUA) and the E 'Alu Pū network of communities for including me in your thoughtful and inspiring work to sustain *'āina* and people.

Mahalo to the many students and community members who helped to gather for this work, including Adam Ayers, Emily Cadiz, Kapua Chandler, Aunty Lahela Chandler Correa, Lucia Olive Hennely, Noah Ka'aumoana-Texeira, Tasia Chase, Lahela Correa Jr., Atta Forrest, Johanna Gomes, Melosa Granda, Elaine Albertson, Lorilani Keohokalole, Amy Markel, Moana and Colin McReynolds, Miki'oi Wichman, Beth Wylie, and the nine students of NREM 620: Kaiāulu, Collaborative Care and Management, spring 2015. Much appreciation to Stephanie Struble, Hui o Hawai'i at Stanford University, Shae Kamaka'ala, Micky Huihui, and Malia Nobrega's LAMA team, for your respectful

Haleleʻa and Koʻolau community members and friends gather to celebrate Makahiki, Limahuli, Hāʻena, November 19, 2017. (Photo by Kilipaki Vaughan)

and careful transcriptions; and to Joel Guy for your video documentation and support. *Mahalo* to Jennifer Luck, partner in this journey who never says never, fellow mother and friend; to Dominique Cordy and Kamealoha Forrest for sharing your expertise with archival land documents and translation, helping me to see things in a new light. I am grateful also for two cousins who have given me the courage and grounding to tell these stories: Billy Kinney, for your insights and wisdom, and Kauʻi Fu, for your strength and truth.

For nudging me to write, I am indebted to The Next Generation Program of Brooke and Terry Tempest Williams, and The Pacific Writer's Connection's Hanalei Writers. Deepest gratitude to my dissertation advisers: Peter Vitousek, Barton Thompson, Margaret Caldwell, Nicole Ardoin, and Louise Fortmann. You each, in your own way, shaped both this work and me through your example as dedicated teachers, discerning scholars, and good people making a difference. *Mahalo* to Pam Matson and Peter Vitousek for bringing me back to school; to Ann and Evan for home away from home and nourishment of all kinds. *Mahalo piha* to Noʻeau Peralto, Debra Gwartney, Johanna Gomes, Sibyl Diver, Mez Baker, Heather Lukacs, Uncle Carlos Andrade, my cousin Noe Goodyear Kaʻōpua, Noenoe Silva, and two anonymous reviewers for taking time to read and enhance this work. Thank you especially to Louise Fortmann, my Mellon Hawaiʻi Fellowship mentor, for reading it, again and again, and telling me to just finish the book. Deepest gratitude to Monica Montgomery, without whose insight, skill, and steady work, it would never be done. Mary Elizabeth Braun, thank you for seeing the potential of this book, shepherding it, and entrusting me to the Oregon State University Press *hui* of Tom Booth, Marty Brown, Micki Reaman, and gifted editor Susan Campbell. All of you taught me with your humble giving, patience, and expertise; a superb birthing team. And *mahalo* to Blaine Namahana Tolentino for birthing the index in the discerning, path-setting way of your *'ohana*. Uncle Gary Smith, your stories, fact checking, and hand-rendered maps enliven our collective knowledge of home, along with this book. Abbey Romanchak, your artistic brilliance and friendship enrich my life and the cover of this book. Anuhea Taniguchi and Mike Coots, you have blessed me with the sharing of photographs only your eyes and love for our community could capture.

Mahalo to my colleagues in the Department of Natural Resources and Environmental Management at UH Mānoa, the College of Tropical Agriculture and Human Resources, UH Sea Grant College Program and CREST, Kamakakūokalani Center for Hawaiian Studies, the EIPER program, indomitable friends of my cohort, and to my *hoa* of Hui 'Āina Momona. It is an honor to work with each of you. *Mahalo* to the funders and contributors who have believed in my work and schooling, including Kohala Center's Mellon-Hawai'i Foundation, Kamehameha Schools, the Office of Hawaiian Affairs, and Deviants from the Norm; National Science Foundation's NSF SEES, UC Berkeley's Community Forestry and Environmental Research Partnerships, and the Switzer and Heinz Foundation; Stanford University's William C. and Jeanne M. Landreth Fellowship, George Rudolf and McGee Fellowships for Summer Research, UH Mānoa's Chancellor's initiatives; and the Hawai'i Community Foundation's Takekiko Hasegawa Scholarship, Ida M. Pope Scholarship, and the North Shore and East Side Kaua'i Scholarship.

Mahalo to the many Kaua'i aunties and friends (including Kari Shozuya, Tasia Chase, Kapua Chandler, Lorilani Keohokalole-Torio, Megan Juran, Megan Wong, Stacy Sproat, Ann Eu, Jody Thayer, Kahanu Keawe, Lei Wann, Kau'i Quinones, and Stephanie Fitzgerald) who have provided child care, airport rides, veggies, poi, meals, and the network of ever-emerging-at-just-the-right-time support that has made this journey possible, you know who you are. *Mahalo* to Aunty Mauli Cook for opening your home, refrigerator, sunny Anahola pasture view, and stream of warm encouragement; to Aunty Carol and Uncle Gaylord Wilcox; and to everyone who let me set up my laptop on your porch. *Mahalo* to Abbey Romanchak, Malia Nobrega-Oliveira, Chia Granda, and Sarah Dryden-Peterson for always being there to talk; to Nā Pi'i Ali'i, Nā 'A'ali'i and all the amazing friends I cannot mention here for inspiring with your presence in my life.

Thank you to all of our family—Vaughan and Blaich clan, and especially to my wise and hardworking mother-in-law, Ipolani Vaughan; and to my Bailey aunties (Karla, Polly, and Fahy), uncles (John, Speedy, Cliff, and Jim), and cousins for keeping us going in so many ways. Thank you to Meleana, for your energy, belief, and being the best sister and aunty, to my steady second brother Will, and nephew Kai'ea. Thank you to my mother for inspiring much of this work through your own and for the most elemental forms of support. Kilipaki, *ho'omanawanui,* thank you for being there and loving me. Pikomanawakūpono, Pi'ina'emalina, and Anaualeikūpuna, *mahalo* for growing with this book and helping your mommy to grow too.

Mahalo ancestors and *akua* for guiding all things, and to all the places and people who have eased, nourished, and inspired this work and taught me along the way. Whether or not you are mentioned, you are not forgotten.

Me ke aloha pumehana ... With warmest aloha

Index

MEHANA BLAICH VAUGHAN grew up where the *moku* (districts) of Halele‘a and Ko‘olau meet on the island of Kaua‘i. She is an Assistant Professor at the University of Hawai‘i at Mānoa in the Department of Natural Resources and Environmental Management, Sea Grant College Program, and Hui ‘Āina Momona. Her home is on Kaua‘i with her husband, mother, and three children. *Kaiāulu* is her first book.